Closing the Loop of Model Predictive Force Control in Milling with Ensemble Kalman Filtering

Schließen des Regelkreises einer modellprädiktiven Kraftregelung beim Fräsen mit einem Ensemble Kalman Filter

Von der Fakultät für Maschinenwesen
der Rheinisch-Westfälischen Technischen Hochschule Aachen
zur Erlangung des akademischen Grades eines
Doktors der Ingenieurwissenschaften
genehmigte Dissertation

vorgelegt von

Max Schwenzer

Berichter/in:

Univ.-Prof. Dr.-Ing. Thomas Bergs
Univ.-Prof. Dr.-Ing. Dirk Abel

Tag der mündlichen Prüfung: 13. Dezember 2021

ERGEBNISSE AUS DER PRODUKTIONSTECHNIK

Max Schwenzer

Closing the Loop of Model Predictive Force Control in Milling with Ensemble Kalman Filtering

Herausgeber:
Prof. Dr.-Ing. T. Bergs
Prof. Dr.-Ing. Dipl.-Wirt. Ing. G. Schuh
Prof. Dr.-Ing. C. Brecher
Prof. Dr.-Ing. R. H. Schmitt

Band 13/2022

WZL | RWTH AACHEN UNIVERSITY

Fraunhofer
IPT

Bibliografische Information der Deutschen Nationalbibliothek
Die Deutsche Nationalbibliothek verzeichnet diese Publikation in der Deutschen Nationalbibliografie; detaillierte bibliografische Daten sind im Internet über https://portal.dnb.de abrufbar.

Max Schwenzer:

Closing the Loop of Model Predictive Force Control in Milling with Ensemble Kalman Filtering

1. Auflage, 2022

Gedruckt auf holz- und säurefreiem Papier, 100% chlorfrei gebleicht.

Apprimus Verlag, Aachen, 2022
Wissenschaftsverlag des Instituts für Industriekommunikation und Fachmedien an der RWTH Aachen
Steinbachstr. 25, 52074 Aachen
Internet: www.apprimus-verlag.de, E-Mail: info@apprimus-verlag.de

Druck: CPI Druckdienstleistungen GmbH
Ferdinand-Jühlke-Straße 7 | 99095 Erfurt

ISBN 978-3-98555-057-9

D 82 (Diss. RWTH Aachen University, 2021)

Preamble

I wrote this thesis at the CHAIR FOR MANUFACTURING TECHNOLOGY at the LABORATORY FOR MACHINE TOOLS AND PRODUCTION ENGINEERING (WZL) OF RWTH AACHEN UNIVERSITY Dezember 2015 – March 2021. I would like to thank PROF. DR. DR. E.H. DR. H.C. DR. H.C. FRITZ KLOCKE and PROF. DR. THOMAS BERGS, MBA as heads of the chair, for their generous support and for providing excellent equipment and working conditions. My special thanks go to PROF. DR. DIRK ABEL, head of the INSTITUTE OF AUTOMATIC CONTROL (IRT), for his expertise and fruitful collaboration with his chair. I would like to thank PROF. DR. KAMKPER, MBA for taking over the chairmanship of the examination and DR. BENJAMIN DÖBBELER and DANIEL SCHRAKNEPPER as my Chief Engineers for the freedom to pursue my ambitions.

The work was funded by the DEUTSCHE FORSCHUNGSGEMEINSCHAFT (DFG) within the Clusters of Excellence "Self-optimizing Production Systems for High-wage Countries" (2006-2018) and "Internet of Production" (2019-2026). It has been my pleasure and honor to work with you brilliant researchers. I sought inspiration and collaboration and found much more. In particular, I would like to thank DR. SEBASTIAN STEMMLER from the IRT for his expertise, inspiration, and dedication. Every time I was on the verge of declaring something impossible after countless attempts, you wiped away all doubts with the spark my motivation needed. MUZAFFER AY, as his successor, you supported me equally well. We made a great team. I would like to thank DR. OLIVER ADAMS for introducing me to it and for providing the idea and the foundation of this multidisciplinary endeavor. I would also like to thank DR. GUISEPPE VISCONTI from the INSTITUTE OF GEOMETRIC AND PRACTICAL MATHEMATICS (IGPM) for patiently explaining the theory of ensemble KALMAN filtering to me.

To all active and former members of the research group "Product and Process Monitoring" and the department of "Cutting Technology", I express my deepest gratitude for your great support, spirit, and an unforgettable time. Special thanks are due to DR. THOMAS AUERBACH for his unwavering trust in my skills and the in-depth discussions that usually took place just before the end of the work day. I want to thank FABIO FILIMBENI for teaching me how to operate a machining center and his help when I failed to do so. Most of all, I want to thank ROBERT SEIDNER and SASCHA KAMPS for their occasional late shifts at the machine and for putting up with my gibberish about programming, run times, dead times, and other nerdy nonsense.

January 2022 Max Schwenzer

Abstract

In milling, the force determines productivity, quality, and safety. Fast-changing engagement conditions between tool and workpiece provoke abrupt changes in the force causing classic, reactive control to fail. Such failure is commonly avoided by using predictive control methods that anticipate process behavior on the basis of process models. One representative, which had been successfully applied to the milling process, is model-based predictive control (MPC).

MPC for force control in milling requires an online identification of a process model. A process model consists of a mechanistic force model and of a model of the radial deviation adjusting the undeformed chip thickness. The state-of-the-art identification method is a repeated curve fit over larger intervals, essentially opening the force control loop in the meantime. The results of these curve fits are uncorrelated and have to be smoothed in order to produce a mathematically continuous feedback signal to avoid instability of the controller. For the identification, the presented work introduces a recursive method based on the idea of a disturbance observer. The observer is a particle filter known as ensemble KALMAN filter (EnKF). EnKF is mathematically proven to converge correctly if the search space is large enough so that it contains the solution. The search space can be narrowed by additional knowledge limiting it to realistic solutions. There exist a large number of experimental reports on mechanistic force models making them well-suited for an EnKF-based identification. The filter states the originality of this work.

After the theoretical background is derived and after the adaption of the filter to the problem case in simulations, the EnKF is studied on real measurements. It shows a remarkable robustness against measurement noise. With regard to online capability, design and setup of the identification, of the controller itself, and of the overall system are put on a proper basis. The system is tested against a conventionally designed process with constant feed, as well as against previous implementations of a model predictive force controller (MPFC) in milling. The contribution to the state of the research are first and foremost robustness as well as the ability to permanently close the feedback loop via the EnKF. The advantage to state-of-the-industry solutions is the reduction of the manufacturing time by 40-60 % compared to a process with constant feed. So far, these contributions do not address a potentially better utilization of the tool life as a result of a more homogeneous load on the tool.

The thesis describes the continuous adaptation of a model in the MPFC and thus enables a truly adaptive, optimal manufacturing process at its technological limit and an increase in workpiece quality.

Kurzfassung

Die Zerspankraft bestimmt beim Fräsprozess Produktivität, geometrische Qualität und Prozesssicherheit. Jedoch sind klassische, reaktive Regelungsverfahren ungeeignet aufgrund der sich schnell ändernden Eingriffsbedingungen am Werkzeug und den daraus resultierenden abrupten Änderungen der Zerspankraft. Prädiktive Regelungsverfahren umgehen dies, indem sie das zukünftige Prozessverhalten modellieren und dadurch vorausschauend agieren.

Die modellbasierte, prädiktive Regelung (MPC) zur Kraftregelung beim Fräsen schließt den Regelkreis durch ein Kraftmodell, welches online identifiziert wird. Dafür stellt diese Arbeit eine rekursive Identifikation mittels Ensemble KALMAN Filter (EnKF) vor, welche auf einem Partikelfilter basiert. Er stellt das Alleinstellungsmerkmal der Arbeit dar. Bisherige Arbeiten nutzen diskrete, unabhängige Identifikationen mittels Kurvenanpassung über größere Intervalle und unterbrechen so den Regelkreis zwischen den einzelnen Identifikationen. Diese Regelungssysteme bilden einen Hybrid aus Kraftregelung und Steuerung.

Es werden die Theorie und die Anpassung des EnKF auf den Problemfall der Identifikation eines Kraftmodells erarbeitet und mit Hilfe von Simulationen analysiert. Danach wird der neue Ansatz anhand realer Messungen validiert. Im Hinblick auf Berechnungseffizienz werden die Identifikation, der Regler selbst und das Gesamtsystem systematisch ausgelegt. Das System wird sowohl gegen einen konventionell ausgelegten Prozess mit konstantem Vorschub, als auch gegen den aktuellen Stand der Forschung einer MPC mit wiederholter Kurvenanpassung getestet. Gegenüber dem Stand der Forschung wird eine deutlich höhere Robustheit erzielt, sowie erstmalig der Regelkreis dauerhaft geschlossen. Gegenüber dem Stand der Industrie mit konstantem Vorschub wird die Fertigungszeit auf 60-40 % reduziert. Die Betrachtung einer potentiell besseren Ausnutzung der Werkzeugstandzeit durch eine homogenere Belastung des Werkzeugs ist nicht Gegenstand der Arbeit.

Die Arbeit beschreibt die kontinuierliche Anpassung des Modells in dem MPC-basierten Kraftregler und ermöglicht dadurch einen adaptiven, optimalen Fertigungsprozess an dessen technologischen Limit.

Contents

List of Figures

List of Tables

Lists of symbols and acronyms

Bold symbols indicate non-scalar variables. Bold lowercase symbols denote vectors, whereas bold uppercase symbols represent matrices. Symbols that are overlined indicate averaged values.

List of uppercase characters

Symbol	Unit	Description
$\boldsymbol{C}^{pp}$		covariance matrix of the model prediction
$\boldsymbol{C}^{up}$		covariance matrix between the states and its prediction
D	mm	diameter of the tool
F	N	process force
F_a	N	active force in the working plane
$\boldsymbol{F}_{act}$	N	actual/ideal force
$F_{a,des}$	N	desired active force, reference for force control; usually the maximum active force per revolution
$F_a\vert_{rev}^{max}$	N	active force, where the measured active force becomes maximum in a revolution
$F_{a,max}$	N	maximum active force
$F_{a,sim}\vert_{rev}^{max}$	N	maximum active force per revolution
F_c	N	cutting force (absolute value), synonym for F_t
F_{cn}	N	cutting normal force, synonym for F_r
F_f	N	feed force, i.e. the force in the direction of the feed velocity
F_p	N	passive force, perpendicular to the working plane
F_r	N	radial force
F_t	N	tangential force (absolute value) generating the cutting torque T_t
$\boldsymbol{F}_{wgn}$	N	noisy force: actual force $\boldsymbol{F}_{act}$ augmented with white GAUSSIAN noise creating artificial measurements
F_{X}	N	force value in the direction of the X-axis of the machine tool
F_x	N	force in a fixed, CARTESIAN coordinate system, perpendicular to F_y
F_{Y}	N	force value in the direction of the Y-axis of the machine tool

Symbol	Unit	Description
F_y	N	force in a fixed, CARTESIAN coordinate system, perpendicular to F_x
$\boldsymbol{G}$	–	gain or KALMAN gain
$G(s)$		transfer function (frequency domain)
$\boldsymbol{H}$	–	measurement matrix; size: $(N_{\boldsymbol{x}} \times N_{\text{obs}})$
$\boldsymbol{I}$	–	identity matrix; ones at the diagonal and zeros at the off-diagonal entries
J	–	ensemble size or number of members of the ensemble respectively
$K_{i,c}$	N/mm	coefficient of the ALTINTAS-model: "cutting coefficient"
$K_{i,e}$	N/mm^2	coefficient of the ALTINTAS-model: "edge coefficient"
K	–	constant gain
$K_{i,m}$		material-specific model coefficient
L	mm	unsupported length of the tool, i.e. the length between tool holder and tip of the tool
$\boldsymbol{M}$		measurement matrix; size: $(N_{\text{obs}} \times N_{\boldsymbol{x}})$
M_X	N m	bending moment of the tool in one fixed direction, caused by the active force
N_{ECsync}	–	number of Δp-steps to be checked to determine the engagement conditions of a certain position of the tool at the tool path
N_1	–	lower prediction horizon of MPC
N_2	–	(upper) prediction horizon of MPC
N_u	–	control horizon of MPC
N_{obs}	–	length of the observation vector $\boldsymbol{z}$, i.e. number of observations
N_s	–	number of disk elements, into which a tool is sliced to approximate its helix angle β as a spiral staircase
$N_{\boldsymbol{x}}$	–	length of the state vector $\boldsymbol{x}$, i.e. number of states
N_t	–	number of teeth of a tool
$\boldsymbol{P}$		model covariance matrix of a KALMAN filter
$\boldsymbol{P}$		covariance matrix (general)
$\overline{\boldsymbol{P}}$		empirical covariance (matrix) of the ensemble of an EnKF
$\boldsymbol{P}_{\boldsymbol{X}0}$		empirical covariance matrix of the initial ensemble of an EnKF; size: $(\boldsymbol{z} \times N_{\text{obs}})$

Symbol	Unit	Description
P_{el}	W	effective electrical power
$\boldsymbol{Q}$	–	model noise covariance of an KALMAN filter
R	mm	radius of the tool
$\overline{\boldsymbol{R}}$		empirical noise covariance matrix; size: $(N_{\text{obs}} \times N_{\text{obs}})$
$\boldsymbol{R}$	–	measurement noise covariance of an KALMAN filter
$T_{95\ \%}$	ms	settling time, in which a system reaches 95 % of the commanded input
T_{cut}	s	cutting time; duration that the tool is engaged in a process
T_d	ms	time delay of the machine tool
T_{prc}	s	process time; duration between the first to the last cut (i.e. from point 1 to 9 in the exemplary workpiece geometry, Fig. 5.3)
T_{rev}	s	time, in which a cutter performs a single revolution
T_s	ms	sample time
T_{all}	s	total time; duration how long the tool is moving
T_t	N m	cutting torque
VB_{max}	µm	maximum width of a flank wear land
$\boldsymbol{W}_u$	–	weight matrix in the cost function of the MPC: change in the manipulated variable
$\boldsymbol{W_r}$	–	weight matrix in the cost function of the MPC: reference tracking
$\boldsymbol{W_\xi}$	–	weight matrix in the cost function of the MPC: slack variables
$\boldsymbol{X}$		array of state vectors; size: $(N_{\boldsymbol{x}} \times J)$
$\boldsymbol{X}_0$		initial ensemble of an EnKF
$\boldsymbol{Y}$		array of model output y for all states $\boldsymbol{X}$; size: $(N_{\text{obs}} \times J)$
$\boldsymbol{Z}$	–	array of (augmented) measurement vectors; size: $(N_{\text{obs}} \times J)$

List of lowercase characters

Symbol	Unit	Description
a_e	mm	width of cut
a_p	mm	depth of cut
b	mm	width of undeformed chip

Symbol	Unit	Description
b_{max}	mm	maximum width of undeformed chip along the entire tool path
Δp	mm	resolution of the tool path
f	mm	feed per revolution
f_t	mm	feed per tooth (and revolution)
$f_{t,max}$	mm	maximum feed per tooth
f_{edge}	Hz	cut-off frequency of a (typically low-pass) signal filter
f_s	Hz	sample rate
h	mm	undeformed chip thickness
h_{crit}	mm	critical undeformed chip thickness, below which ploughing occurs
h_{max}	mm	maximum undeformed chip thickness per revolution
h_{def}	mm	deformed chip thickness after cutting
k_i	$\mathrm{N/mm^2}$	gains of the KIENZLE-model
k_r	$\mathrm{N/mm^2}$	radial gain of the KIENZLE-model
k_t	$\mathrm{N/mm^2}$	tangential gain of the KIENZLE-model
k	–	discrete time step
k'	–	corrected discrete time step
k^*	–	ensemble inflation: step size of ensemble inflation
m_i	–	exponents of the KIENZLE-model
m_r	–	radial exponent of the KIENZLE-model
m_t	–	tangential exponent of the KIENZLE-model
n	1/s or 1/min	rotation speed
r_β	°	rounded cutting edge radius
$\boldsymbol{r}$		reference or desired value(s)
u		command, input, or control action
$\boldsymbol{u}$		input vector
v_c	m/s	cutting velocity
v_f	mm/s	feed velocity
$v_{f,act}$	mm/s	actual/measured feed velocity
$v_{f,cmd}$	mm/s	commanded feed velocity
$v_{f,des}$	mm/s	desired feed velocity
$v_{f,KF}$	mm/s	filtered feed velocity. This is the effective value of two separate KALMAN filters for the X- and Y-axes of the machine tool
$v_{f,max}$	mm/s	maximum allowed feed velocity of the machine tool or within the process

Symbol	Unit	Description
$\boldsymbol{x}$		state vector; size: $(N_{\boldsymbol{x}} \times 1)$
$\boldsymbol{y}$		output vector of a model
y		output of a model / control variable
z	mm	height, e.g. along the major axis of a tool
$\boldsymbol{z}$		measurements; size: $(N_{\text{obs}} \times 1)$

List of Greek characters

Symbol	Unit	Description
α_0	°	tool orthogonal clearance or radial primary clearance; angle between the flank face and the tangential direction
α_f	°	feed direction angle
$\alpha_{\overline{\boldsymbol{P}}}$	–	inflation factor for variance inflation
β	°	helix angle of a tool
$\Delta\varphi_\beta$	°	circular sector that a cutting edge spans due to its helix angle β
$\boldsymbol{\eta}$		noise vector; here: white, GAUSSIAN noise; size: $(N_{\text{obs}} \times 1)$
γ_0	°	rake angle of the tool
$\boldsymbol{\Gamma}$		noise covariance matrix; size: $(\boldsymbol{z} \times N_{\text{obs}})$
κ	–	forgetting factor in the RLS: $\kappa \in (0, 1)$
λ^*	–	ensemble inflation: fraction of the initial covariance $\boldsymbol{P}_{\boldsymbol{X}0}$ to which the ensemble is inflated
λ	°	radial deviation: direction of the radial offset (KLINE & DEVOR-model)
ω_0	Hz	cut-off frequency
φ	°	angular position of the spindle
φ_t	°	angular position of the cutting edge
${\varphi_t}^\star$	°	angular position of the cutting edge, where the undeformed chip thickness $h(\varphi_t)$ becomes maximum
φ_{ex}	°	end or exit angle of an engagement
φ_{in}	°	start angle of an engagement
Φ	°	shear angle defined between feed direction and shear plane
$\phi(n)$	–	BOOLEAN function denoting whether a cutting edge is engaged
ρ	µm	radial deviation: radial offset (KLINE & DEVOR-model)

Symbol	Unit	Description
ρ_i	µm	radial deviation: radial deviation parameters in CARTESIAN coordinates
ρ_x	µm	radial deviation: radial offset in CARTESIAN coordinates
ρ_y	µm	radial deviation: radial offset in CARTESIAN coordinates
σ		standard deviation
τ	°	radial deviation: tilt of the tool (KLINE & DEVOR-model)
ψ	°	radial deviation: direction of the tilt of the tool (KLINE & DEVOR-model)
$\boldsymbol{\xi}$		slack variable(s) of an MPC
ζ	–	damping ratio

List of acronyms

cRIO	compact reconfigurable input/output
S/N	signal-to-noise ratio
i.i.d.	independent and identically distributed
s.t.	subject to
A/D	analog to digital
ACC	adaptive control with constraints
ACO	adaptive control with optimization
ADAS	AXIS DATA STREAM
API	application programming interface
BOBYQA	BOUNDED OPTIMIZATION BY QUADRATIC APPROXIMATION
CAM	computer aided manufacturing
CIRP	COLLEGE INTERNATIONAL POUR LA RECHERCHE EN PRODUCTIQUE
CPU	central processing unit
CSG	constructive solid geometry
CSV	comma separated value
DDR	double data rate
DEnKF	dual ensemble KALMAN filter
DKF	dual KALMAN filter
DLL	dynamic-link library
EnKF	ensemble KALMAN filter
FEM	finite element method
FFT	fast FOURIER transformation

FPGA	field programmable gate array
GAC	geometric adaptive control
GCC	GNU COMPILER COLLECTION
GPC	general predictive control
IEEE	INSTITUTE OF ELECTRICAL AND ELECTRONICS ENGINEERS
ILC	iterative learning control
KF	KALMAN filter
KKT	KARUSH-KUHN-TUCKER
LAN	local area network
MFT	mean field theory
MinGW	MINIMALIST GNU FOR WINDOWS
MPC	model-based predictive control
MPFC	model predictive force controller
MRAC	model reference adaptive control
MRR	material removal rate
NC	numerical controller
NCU	numerical control unit
NEWUOA	NEW UNCONSTRAINED OPTIMIZATION ALGORITHM
NLopt	NONLINEAR OPTIMIZATION
NN	neural network
ODE	ordinary differential equation
qpOASES	QUADRATIC PROGRAMMING ONLINE ACTIVE SET STRATEGY
RAM	random-access memory
RCD	rotating cutting dynamometer
RLS	recursive least squares
rms	root mean square
SN	serial number
SQP	sequential quadratic programming
SVR	support vector regression
TCP	tool center point
TSN	time-sensitive network
TTL	transistor-transistor logic
UDP	user datagram protocol
UKF	unscented KALMAN filter

1 Introduction

Milling is a flexible, economical, and well-established manufacturing process. It is characterized by a cyclically intermittent cut with a defined edge. The process force is the major parameter that influences productivity, quality, and safety of the process. A high mechanical tool load usually signifies a high material removal rate (MRR) favoring productivity, but it also deflects the tool and potentially the workpiece. This deflection may result in a violation of geometric tolerances of the workpiece, which correspond to a quality measure. Tool wear is driven by the thermo-mechanical load, to whose cyclical alternation brittle materials react particularly sensitively. Large peaks in the force may boost tool wear and provoke tool failure putting process safety at risk. Almost $20\,\%$ of the downtime of a machine tool is due to tool failure [LIAN04].

Controlling the force induces a paradigm shift in the design of the milling process. Although research on force control has been conducted since the beginnings of metal cutting, it has been restricted to laboratory work. The reason for this limited application was that the reactive nature of classic controllers could not prevent force peaks in the presence of abrupt changes in tool engagement, e.g. when the tool enters the workpiece. The great advances in construction and design of machine tools and their controllers cannot hide the fact that machine tools control setpoints, such as position, rather than process variables, such as force or temperature [THOM18]. Machine tool variables are only indirectly related to the process and to the resulting product quality. Process variables, such as force, are directly coupled to the targeted process outcome.

Already in 1993, Ulsoy & Koren [ULSO93] acknowledged the economic impact of process control by a back-of-the-envelope calculation. Even small increases in the productivity cumulate dramatically over the lifetime of a machine tool. (They suggested a return of invest of 1000 times.) For force control, they identified that: *"The basic problem arises from the fact that the effective process gains and time constants depend upon process variables such as feed, speed, and depth of cut. Furthermore, process related parameters such as tool geometry, work, and tool material properties are difficult to characterize and can have a significant effect on the process dynamics. This unpredictability of the process can lead to poor performance, tool breakage, or even instability when fixed gain process controllers are used"* [ULSO93, p. 304]. The industrial answer to these problems was a simulation-based process design. Such simulations helped compensate known disturbances, but not uncertainties such as tool wear. Thus, the process design have to be conducted for a hypothetical worn tool in order to still meet the quality requirements of the workpiece at the end of the tool life.

Advanced control methods address the limitations of classic controllers fusing simulation and control theory. MPC is one representative of this class of novel methods. It explicitly considers a process model, as well as technological constraints, to determine a new optimal control law at every time step. The previous works of ADAMS [ADAM16a] and STEMMLER [STEM20] at the RWTH AACHEN UNIVERSITY demonstrated its applicability to the milling process. Using MPC to control the force enables, in theory, a safe, overshooting-free, optimal control at the technological limit of the milling process. It homogenizes predominantly the mechanical tool load keeping it in a desired range in spite of fast changing engagement conditions.

ULSOY & KOREN [ULSO89] have claimed that a successful application of a process control is hindered by the lack of a reliable tool wear model and monitoring. This work follows the hypotheses that the effect of tool wear manifests itself in the process force. Therefore, controlling the force inherently controls the tool wear rate and additionally avoids tool breakage while optimizing productivity. In fact, the argumentation is that a model predictive force controller (MPFC) for milling:

- guarantees geometric product quality by limiting the active force, which causes deflection of the tool and thus, may violate specified workpiece tolerances,
- ensures process safety by avoiding tool overload, which ultimately causes failure of the tool,
- would improve profitability by prolonging the tool life through a reduction of the load variance, while
- maximizing productivity by optimizing the feed resulting in a minimal production time.

The presented work addresses the question of how to provide a continuous adaption of the process model, on which the MPFC relies. This adaption closes the feedback loop of the force. The quantification of the economic impact is left for further research.

2 State of the art

This chapter first discusses different approaches of force control for milling (Section 2.1) with a focus on predictive approaches (Section 2.1.5). Later, process models are reviewed together with their calibration – called "identification" (Section 2.2). Here, a process model denotes the combination of a force model and a model of the radial deviation of the tool. Special attention is paid to the online capability of the identification approaches. A distilled overview provide Tab. 2.1 and Tab. 2.4, before Section 2.3 summarizes the discussion.

2.1 Force control in milling

Free-form surfaces in milling lead to fast-changing engagement conditions of the tool, which in turn cause a varying cutting force. The load on the cutting edge (cutting force and cutting normal force) significantly affects the tool life [IBAR10]. The active force F_a – the effective value of cutting and cutting normal force – induces a bending of the tool. Thus, it is crucial for tool breakage and tool deflection. The tool deflection again determines the dimensional accuracy of the manufacturing process and, thereby, one aspect of product quality.

A note to clarify the terms in this work: In the Anglo-American literature "cutting force" is often misleadingly used as a synonym for "active force" (e.g. [ALTI90; NOLZ95a; ROBE96; DENK12; ALTI17]). In some cases, the term "resultant force" is used too [LAUD89; CUS11; LAYE12], which sometimes also denotes the norm of the three-dimensional force vector. This work follows the CIRP DICTIONARY OF PRODUCTION ENGINEERING [INTE97], explicitly distinguishing the terms cutting force $F_c \equiv F_t$ (also known as tangential force) and cutting normal force $F_{cn} \equiv F_r$ (or radial force), which compose the active force F_a in the working plane as CARTESIAN vectors:

$$\vec{F_a} = \vec{F_c} + \vec{F_{cn}} \equiv \vec{F_t} + \vec{F_r}. \tag{2.1}$$

The terms tangential force F_t, radial force F_r, and active force F_a are used in the following to avoid any confusion between cutting force and active force.

"Process control, which is not commonly integrated in today's machine tools, is the automatic adjustment of process parameters (e.g. feeds, speeds) in order to increase operation productivity and part quality" [LIAN04, p. 297]. Controlling process parameters rather than machine tool setpoints may even result in a closed-loop (workpiece-)quality control [THOM18], what ULSOY & KOREN [ULSO93] named "supervisory control". Addressing the overall

economic optimization of the machining process, they foretold an in-process quality assurance.

Research in the field of process control for milling has been overwhelmingly concentrated on force control since it affects almost all other phenomena [LIAN04]. Force control in milling contributes to productivity (by faster manufacturing), safety (by avoiding tool breakage), and profitability (by prolonging tool life) while maintaining product quality. Hence, force control in milling is under constant research since the late 1970s, e.g. [STUT76; SCHE79; KAPP79; KORE81; BRAN81]. The relevant literature presented here is restricted to digital controllers.
The former classification of adaptive control with constraints (ACC), adaptive control with optimization (ACO), and geometric adaptive control (GAC) has outlived. ACC holds a process variable within given bounds, ACO optimizes a performance index, and GACs *"seek to maximize the quality of finishing operations in the face of structural deflections and tool wear"* [LIAN04, p. 303]. The recent introduction of MPC to force control combines reference tracking with constraints, which will be discussed in Section 2.1.5. This work refers to "constrained control" if the control variable and/or the manipulated variable can be constrained (e.g. due to actuator saturation). IJAZ ET AL. [IJAZ14] pointed out that a constrained control of the tool load as the control variable ensures high productivity in milling while minimizing the risk of tool breakage. In the beginnings, constraint control was driven by practical considerations and implemented basically by switching the control law or gain if the constraint was violated, e.g. [SCHE79; KORE81]. The term "optimal control" should apply to systems that track a reference of the control variable that is optimal in a given case – e.g. maximum force yields to maximum productivity, or operating at a defined force level enhances tool life resulting in an optimal process in terms of economics [SCHE79, p. 16].

Almost exclusively the manipulated variable u is the feedrate f, with a few exceptions, e.g. [DENK12] (radial engagement), or [BRAN81] (cutting speed, but controlling tool wear instead of force). From control and technological perspective, the feedrate f is the most important control variable regarding the load of the tool, machine, and workpiece [SCHE79, p. 21], and *"among in-process machining parameters, feedrate is the easiest to manage"* [MATS09, p. 448]. It directly affects the process force through the undeformed geometry of the chip, Section 2.2. As control variable y, predominantly the maximum active force is used $y = F_{a,max}$. Tab. 2.1 provides an overview of the control variables and the manipulated variables of selected works.

Many works use the rotation speed n as the sampling rate for the controller: $T_s \in [38\,\text{ms}, 116\,\text{ms}]$ [ALTI90; ALTI94]. ADAMS ET AL. [ADAM16b] used half of

Table 2.1: *Overview of the controlled and manipulated variables for force control in milling. The control method is marked by superscripts.*

Control variable		References
Maximum active force	$F_{a,max}$	[LAUD89]*‡ [SPEN91]$'^{\star}$ [YAMA91]$'^{\star}$ [FUSS91]* [ALTI94]‡ [RICH02]$'^{\star\ddagger}$ [CUS06]$'^{\star}$ [ZUPE06a]$'^{\star}$ [XU08]‡ [CUS11]* [LAYE12]$'$ [ZUPE12b]$'^{\star}$ [NOUR15]$'^{\star}$ [ADAM16b]‡ [STEM16]‡ [ALTI17]‡ [ZHAN18b]$'$ [PARK18]$'$ [STEM20]‡
Mean active force	$\overline{F_a}$	[DENK12]*
3D resultant force	$\|\vec{F}\|$	([ERDI07])$'$ ([KARU09])$'$
Feed force	F_f	[NOLZ95b]*‡ [KIM96]* [ROBE96]‡ [ROBE97]*
Cutting force/ torque	F_t	[KAPP79]*
Bending moment	M_X	[TU04]*
Effective power	P_{el}	[SCHE79]* [NOLZ95a]‡ [TU14]*

manipulated variable: ____ not feedrate, v_f otherwise
control method: $'$ *a priori* optimization, * time-invariant control, ‡ time-variant control

the delay T_d of the manipulating actuator as the sampling time – the actuator are the auxiliary drives in case of the feedrate. Reacting faster than the system dynamics lets the controller better exploit its potential.

For technical systems, it is crucial whether the control system can deal with inherent limitations, such as actuator saturation, system delay, and its usability regarding the tuning of the controller. ULSOY & KOREN [ULSO89] and LIANG ET AL. [LIAN04] drew the same conclusion although both review articles were separated by 15 years of research: In spite of the tremendous effort, process control had only little impact on the industry in manufacturing. MATSUBARA & IBARAKI [MATS09] gave three reasons for this disillusioning result:

- difficult integration,
- sensor (installation) cost, and
- lack of demonstrating the benefit to the operator.

NOLZEN [NOLZ97] and LIANG ET AL. [LIAN04] added to this list the lack of reliability and of robustness.

The different approaches can be clustered into five categories, which will be discussed in more detail in the remaining part of this section:

- *a priori* process optimization,

- time-invariant control (fixed-gain controllers),
- adaptive control (time-variant),
- simulation-enhanced or hybrid control, and
- model-based predictive control (advanced, time-variant control).

The first bullet represents an open loop or feedforward control (Fig. 2.1a) with help of a process simulation in the design phase of a manufacturing process. The remaining bullets classify different methods of closed-loop control (Fig. 2.1b). Simulation-enhanced control presents a special classification of controllers for the milling process.

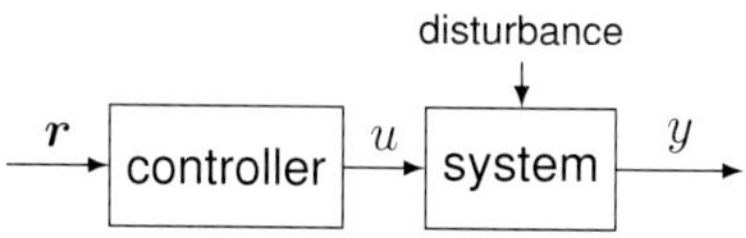

(a) *open loop*

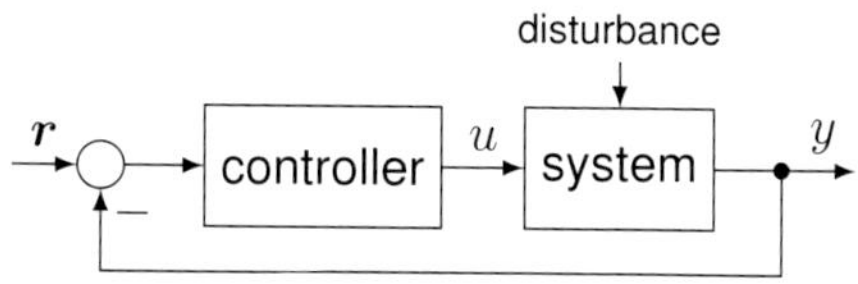

(b) *closed loop*

Figure 2.1: *Open-loop control vs. closed-loop control*

2.1.1 *A priori* process optimization: feedrate scheduling

Feedrate scheduling describes an open-loop control to optimize either the material removal rate (MRR) (volume-based) or the tool load (force-based) [KURT11]. Both serve as a measure of productivity, whereas the force is rather an indirect but physically more justified representation of productivity.
Volume-based process design is state of the industry. Most commercial computer aided manufacturing (CAM) systems aim at generating a tool path with a constantly high MRR at constant feed, e.g. the ADAPTIVE ROUGHING software module by MODULWORKS [MODU21], which is part of many CAM products. In contrast to these constant-feed approaches, a volume-based feedrate scheduling is available at SOLIDCAM under the product name IMACHINING [SOLI21]. The competitor CGTECH as well offers a feedrate scheduling in its product VERICUT OPTIPATH [CGTE21].

Recently, the industry moves towards force-based approaches catching-up with the state of the research. Force-based approaches are generally more precise but more complex in the underlying simulation because they require a force model in addition to engagement conditions [KURT11]. ERDIM ET AL. [ERDI07] showed that a force-based feedrate scheduling was able to reduce the process time by 63 % compared to a process with a constant feed. Compared to a volume-based feedrate scheduling, the latter came at the cost of worse surface quality. The higher feed per tooth in the volume-based approach resulted in distinct marks from the teeth on the surface. The reason for the higher feed was that the force increased disproportionately at small material volumes leading to a lower feedrate in the force-based schedule. PARK ET AL. [PARK18] scheduled the feedrate using a linear relationship from the MRR to the spindle power. Assuming a fixed ratio of tangential and radial force components $F_r = 0.5\ F_t$, they extended the model to the active force F_a:

$$F_a = \sqrt{{F_t}^2 + {F_r}^2} = \frac{\sqrt{5}}{2}\ F_t = \frac{\sqrt{5}}{2}\ \frac{1000}{\pi\ D\ n}\ \underbrace{K_{i,m}\ MRR}_{\text{spindle power}}, \tag{2.2}$$

with the tool diameter D, the revolutionary speed n of the spindle (in rpm), and a material-specific parameter $K_{i,m}$. The material removal rate (MRR) was proportional to feed f and constituted the optimization variable.

Other researchers suggested building an optimization of all process parameters (feed, cutting speed, depth of cut, width of cut) integrating an MRR-simulation with a force model and a simple model for the tool life [MA19]. They searched for an economic optimum using global optimization of the manufacturing cost as a function of the remaining tool life, the manufacturing time, and the setup time. In the presented case, the optimization took 4 h to reduce the manufacturing time by 25 % and the cost by 7 % compared to a manually designed process.

MATSUBARA & IBARAKI [MATS09] combined feedrate scheduling with monitoring of the process force and, with it, indirectly the wear status of the tool. The schedule was repeatedly updated at intervals of *"several dozen meters"* – a detailed example suggested an update every 25 m [IBAR10]. This check-point-based control was of closed-loop type but with a sampling time so large that no short-term process effects were covered.

2.1.2 Time-invariant control

Time-invariant control denotes a closed-loop controller whose behavior does not change, i.e. they have a fixed control law, Fig. 2.1b. Time-invariant control includes classic PID control, where the different letters represent the behavior of the gain: proportional (P), integral (I), and differential gain (D).

KAPPEN [KAPP79] was among the first to implement a digital controller in milling. Two strain gauges at the spindle enabled him to control the cutting torque $T_t \propto F_t$ with a PI controller. The effort was high to cope with the limited computational resources in the early beginnings of microcontrollers but the results demonstrated feasibility. The controller was able to reduce the deviation from a desired torque in the presence of a step in the depth of cut a_p (slot milling: $D = 125\,\mathrm{mm}$, $n = 315\,\mathrm{rpm}$, step height $\Delta a_p = 2\,\mathrm{mm}$ [KAPP79, p. 120]). SCHENKE [SCHE79] presented a limit control in face milling. The system increases the feedrate as long as the spindle power and the displacement of the spindle remained in a certain range. Such type of control (in this case with a P gain) was popular in the beginning when the design was driven by feasibility and not by pole placement in the LAPLACIAN domain. Commercial control systems for milling today still work with similar strategies, also to avoid having to redesign this for each machine tool. They too control the effective power of the spindle, e.g. TOOLSCOPE ADAPTIVE FEED CONTROL (CERATIZIT S.A.) [CERA21], ARTIS ADAPTIVE CONTROL (MARPOSS MONITORING SOLUTIONS GMBH) [MARP21], or TOOLINSPECT ADAPTIVE FEED CONTROL (MCU GMBH) [MCU 21]. These products are time-invariant controllers – although misleadingly promoted as "adaptive controllers" arguing that they adapt the feedrate. Even academic works are not immune to this mistake as ULSOY & KOREN [ULSO89] criticized (further examples are [BRAN81; CARR97; ZUPE06b; CUS06]). Nevertheless, research went on examining how to avoid massive overshoots at abrupt changes in the engagement conditions.

Spindle power is an easy measure and control became more tempting with the upcoming of high-speed machining in the 1980s. So that THE BOEING COMPANY looked into the control of milling as TU & CORLESS [TU14] later reported – one of the rare reports on industrial application. They emphasized that time-invariant controllers reduced spindle failure by $90\,\%$. They examined different controller designs and control variables: robust control to take account of the time delay and the bending moment at the spindle as a control variable [TU04]. NOLZEN & ISERMANN [NOLZ95b] performed an actual force control (on the active force) describing the relationship between feed and maximum active force by a (linearized) exponential force model (see [KIEN52], Eq. (2.7)).

Only recently, ZHANG ET AL. [ZHAN18b] adjusted the feedrate offline between repetitive cuts based on force measurements of the previous cycle performing a constraint optimization. Although this approach is time-invariant within each cycle, it acknowledged that a constant feedrate is not optimal for the milling process. Essentially, it is a repetitive feedrate scheduling but based on (force) measurements providing a feedback loop and turning it into closed-loop control across individual cycles. Such types of control are known under the term iterative learning control (ILC).

2.1.3 Time-variant control: adaptive controllers

The complexity and dynamics of the milling process pose a challenge on an accurate design of time-invariant controllers. Robust controllers may better cope with the fast dynamics, but always assume the worst-case [ABEL06, p. 234]. *"[The] unpredictability of the process can lead to poor performance, tool breakage, or even instability when fixed gain process controllers are used. ... process model uncertainty, were a primary reason for the poor industrial acceptance of the first generation of AC [adaptive control] systems for machining. To address this problem, adaptive control schemes, combining online parameter estimation and control, were developed"* [ULSO93, p. 304]. In adaptive control, the controller parameters were updated online. In this way, it adjusted itself to changed working points or environmental conditions. The characteristic of this control was a system identification – often through an artificial excitation – and the adjustment of the control parameters [ABEL06, p. 175]. An example in cutting technology was given by WIRTZ [WIRT10], where the feed of a gun drilling process was superposed with an artificial sine wave to provide the required excitation for the identification of the system. The adaptive controller was able to eliminate this excitement so that it did not affect manufacturing but enabled for continuous identification of the process behavior.

Adaptive control can be distinguished between approaches without and with a reference model, Fig. 2.2. The former adjusts the parameters solely based on the control system accuracy, while the latter tracks a modeled reference $\hat{y}$: model reference adaptive control (MRAC). Designing the controller itself requires less modeling effort, but inherent inaccuracies of the system may lead to an incorrect tuning of the controller. Early approaches relied on "gain scheduling", which adjusted the gain of a PID controller depending on the current operating point. ALTINTAS & MA [ALTI90] developed an adaptive controller for the maximum active force $F_{a,max}$ in milling. They used a simple nonlinear model of the maximum force (per revolution) and the feedrate based on a static deformation model. Simulation and experiments showed good results for a slowly changing ramp in the depth of cut but overshot the desired force level by $80\,\%$ at sudden changes. NOLZEN & ISERMANN [NOLZ95b] were able to eliminate the constant offset of a P controller by adapting the controller gain. This was exactly the initial motivation of why to use adaptive control in milling. The adaptive controller exceeded the reference force by $50\,\%$ (100 N) for less than a cutter revolution. They also examined effective power as a control variable [NOLZ95a; NOLZ97], but the experiments indicated that the effective power had a certain delay to the force. This observation suggested that the peak cutting force would have been even higher and that an actual force control was required for a safe operation. FUSSELL & SRINIVASAN [FUSS91] could not confirm a better transient behavior of adaptive control in milling than time-invariant control but concluded that adaptive control was more stable over a wider range of cutting conditions. Their simulations stressed that the radial deviation of the tool induced oscillations, which were hard to control (see [STUT76]). As a consequence, more advanced techniques, such as MRAC, were examined.

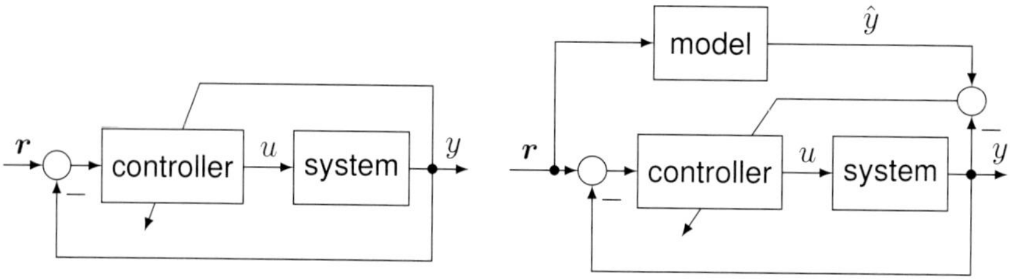

(a) *adaptive control* ***(b)*** *MRAC*

Figure 2.2: *Sketch of (model reference) adaptive control loops*

In MRAC, i.e. adaptive control with a reference model, the controller adjusts its parameters based on the difference between an identification of the closed-loop system and a reference model describing the desired system behavior, Fig. 2.2b. The control parameters are varied until the behavior of the control loop matches the reference model. ULSOY & KOREN [ULSO89] listed examples for milling. MRAC improved the transient behavior in the occurrence of a change in depth of cut [LAUD89]. But it suffered from poor convergence in the case of a constant reference – compared to a PI controller. Measurement noise further reduced the performance. So it was concluded that both, the measurement noise for the identification and the design complexity for the controller, limited an industrial application. The quality of control depended on the model accuracy and measurement noise [ULSO89]. Signal filters deteriorated the reaction of the controller and, with it, its quality particularly at abrupt changes in the engagement conditions. FUSSELL & SRINIVASAN [FUSS91] observed a smoother control behavior of MRAC. It could compensate oscillations induced by the radial deviation of the tool outperforming time-invariant and adaptive control in terms of stability and convergence.

NOLZEN [NOLZ97] noted that adaptive control systems had not been reliable enough for broad success in the industry. ULSOY & KOREN [ULSO89] were not aware of any successful commercial application of an adaptive control system in the industry. The primary reason was the limited reaction time [CUS11], which was inherent in terms of inertia [PARK04]. Adaptive control can stabilize process setpoints at slowly varying process conditions. Hence, it is ideal for cutting processes, where the transfer behavior changes slowly – as drilling [WIRT10] or turning [DANE86]. The reactive nature of all classic controllers (be it time-invariant or time-variant) is incapable of preventing force peaks caused by abrupt changes in workpiece geometry. Sudden changes in the engagement conditions cause large overshoots of the control variable (e.g. [NOLZ95b]: by 50 %, [ALTI90]: by 80 %, [ALTI94]: by 40 %). *"Any effective control scheme must be able to anticipate geometry changes"* [RICH02, p. 182].

2.1.4 Simulation-enhanced control

TU ET AL. [TU04] ascribed the unsatisfactory performance of classic controllers to the unknown cross section of the chip. The issue of low performance promised to be solved by simulation-enhanced control approaches. Just after the occurrence of adaptive control approaches, first ideas were persuaded to combine (offline) simulation of the process – predominantly engagement simulations in combination with a force model – with control methods to achieve a better overall force control for the milling process. While the previously categorized MRAC minimized the difference between a system and a model, the controllers here used a process simulation as a separate source of information without feedback.
SPENCE & ALTINTAS [SPEN91] superposed the measured force with predicted force from simulation letting the controller react to the future process behavior. This look ahead eliminated severe force overshoots in spite of steps in the depth of cut. However, the constant overlay of the estimated force yielded a suboptimal overall control performance. Instead of manipulating the feedback, YAMAZAKI ET AL. [YAMA91] performed an offline simulation of the milling process to generate the optimal reference in terms of MRR. This reference feedrate was then implemented by a time-invariant controller. RICHARDS ET AL. [RICH02] picked up this concept in 2002 arguing that the controller compensated for errors and inaccuracies of the model that was used to generate the optimal reference. They studied a P controller and an adaptive controller confirming previous findings: both controllers showed nearly identical performance but the P controller reacted faster to changes while the adaptive controller provided smoother feedrate commands. Later, ZUPERL & CUS [ZUPE12b] and ZUPERL ET AL. [ZUPE12a] combined an offline feedrate optimization with a neural-network-based controller to prevent overshoots of the force at abrupt changes in depth of cut. Further, in this combination, they could compensate a constant offset, which resulted from model inaccuracies if the scheduler would

be used alone. The neural network (NN) mimicked a division controller (P gain).

TUKORA & SZALAY [TUKO11] suggested to combine an online process simulation with an online adapted force model to adjust the feedrate. The idea remained a vision as the demonstration was performed offline. Only recently, ALTINTAS & ASLAN [ALTI17] realized the idea of identifying the force model online in combination with an adaptive controller. A one-second look ahead was used in the model to anticipate changes in the engagement conditions. KLUGE [KLUG99] went one step further, outlining MPC for metal cutting but discarded it for the metrological effort. Instead, he used a time-invariant controller and stiffness models of the tool and workpiece to compensate static deformations in the turning process to improve the workpiece quality.

2.1.5 Time-variant control: Model-based predictive control

General theory

Model-based predictive control (MPC) is a set of advanced control methods, which uses a system model explicitly to predict the future behavior of the system and to optimize it accordingly, Fig. 2.3. It solves a constrained optimization problem at every sampling point: adjusting the manipulated variable u in such a way that the control variable y best fits the desired variable $\boldsymbol{r}$ over the considered future horizon N_2. The double-lined arrows in Fig. 2.3 indicate that the problem is solved for multiple time steps (i.e. the horizon), so that the variables become vectors of time series. Only the first value of the resulting series of manipulated variables u is implemented. This is why MPC is also referred to as "receding horizon" control. The optimization is the main difference from conventional control with a precomputed control law [MAYN00]. MPC can even control systems, for which the design of a standard controller is impossible and thus cannot be controlled by conventional feedback controllers [RAWL00; MAYN00].

Assuming an arbitrary system with a model function $\boldsymbol{f}(\cdot)$ and with the measurement function $\boldsymbol{h}(\cdot)$:

$$\boldsymbol{x}(k+1) = \boldsymbol{f}(\boldsymbol{x}(k), u(k)), \tag{2.3}$$

$$y(k) = \boldsymbol{h}(\boldsymbol{x}(k)). \tag{2.4}$$

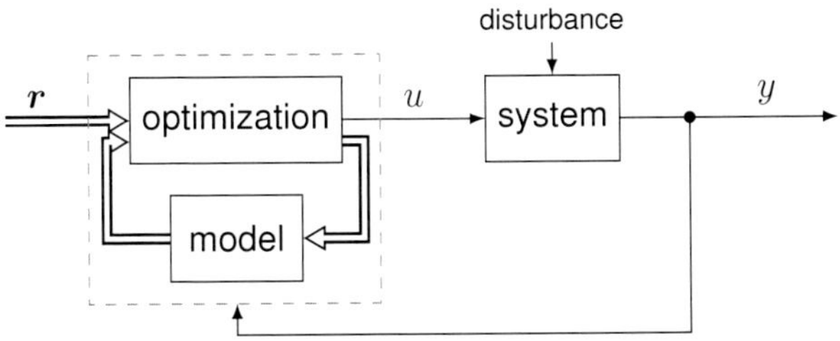

Figure 2.3: *Sketch of a model predictive controller in analogy to Fig. 2.1 and Fig. 2.2. The double-lined arrows indicate that the optimization is conducted for multiple time steps (for the control horizon N_u)*

MPC solves an open-loop optimal control problem at every time instant k. One common formulation of this problem is, [STEM20, p. 82]:

$$\begin{aligned} \min_{u,\boldsymbol{\xi}} \sum_{i=N_1}^{N_2} &\|\boldsymbol{r}(k+i|k) - y(k+i|k)\|^2_{\boldsymbol{W}_r} + \\ &\sum_{j=1}^{N_u-1} \|\Delta u(k+j|k)\|^2_{\boldsymbol{W}_u} + \sum_{i=N_1}^{N_2} \underbrace{\|\boldsymbol{\xi}(k+i|k)\|^2_{\boldsymbol{W}_{\boldsymbol{\xi}}}}_{\text{softening}}, \qquad (2.5) \\ \text{s.t.} \qquad & u_{lb} \leqslant u(k+j|k) \leqslant u_{ub}, \\ y_{lb} - \boldsymbol{\xi}(k+i|k) &\leqslant y(k+i|k) \leqslant y_{ub} + \boldsymbol{\xi}(k+i|k), \quad \text{where } \boldsymbol{\xi} \geqslant 0, \\ &\forall i \in \{N_1, \cdots, N_2\} \quad \text{and} \quad j \in \{0, \cdots, N_u\}. \end{aligned}$$

Bold typed variables indicate higher dimensions, i.e. a vector or a matrix. The predicted state $k+i$ at the discrete time step k is indicated as $\boldsymbol{x}(k+i|k)$. The absolute time t can be reconstructed with the sample time T_s: $t = k\,T_s$. Note that the minimization is conducted over the whole prediction horizon and control horizon, i.e. for all values of i and j in Eq. (2.5).

The cost function minimizes the deviation from the reference $\boldsymbol{r}$ over the prediction horizon N_2. It additionally considers the change in the manipulated variable $\Delta u_k = u_k - u_{k-1}$. The last term includes slack variables $\boldsymbol{\xi}$, which quantify the violation of output constraints. Each term posses an individual weighting matrix $\boldsymbol{W}$. The quadratic norm can be resolved to a matrix multiplication: $\|\boldsymbol{x}\|^2_{\boldsymbol{W}} = \boldsymbol{x}^\intercal\,\boldsymbol{W}\,\boldsymbol{x}$.

Introducing slack variables $\boldsymbol{\xi}(k+i|k)$ relaxes the hard constraints on the output y of the model ensuring feasibility of the optimization problem. It creates an extra degree of freedom, which is penalized in the cost function through the weight W_ξ. The slack variables ξ do not resemble a function but are an individual series for every time step k. Fig. 2.4 illustrates the function idea and the variables. Details on its applications, stability theory, and practical considerations can be taken from [SCHW21].

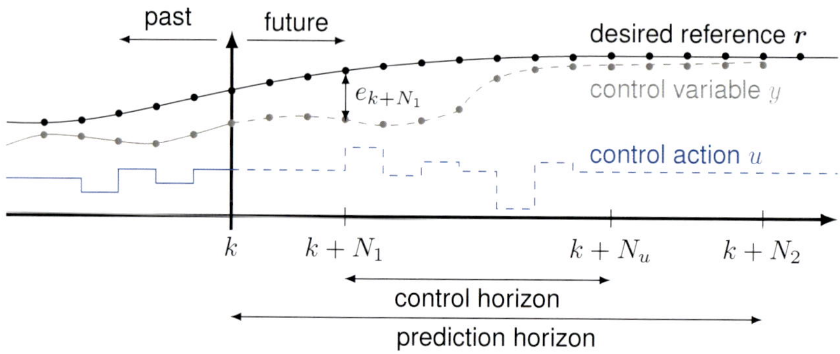

Figure 2.4: *Function principle of MPC indicating the horizons N_1, N_2, N_u (in accordance with [RICH78])*

Model-based predictive control for force control in milling

While MPC has been examined for the position control loop of machining centers [STEP11; BERN19], the work of ALTINTAS & ASLAN [ALTI17] on process control already reflected characteristics of general predictive control (GPC), a simplified flavor of MPC from its early beginnings. They used a linear model and short horizons ($N_2 = 4$). ADAMS and STEMMLER [ADAM16a; STEM16; SCHW17; STEM20] were the first to implement MPC to control the active force in rough milling in an MPFC. Their approach was to model the milling process with two separate models: one for the behavior of the machine tool and one for the cutting process, Fig. 2.5. This was in accordance with other works, such as [FUSS91; ALTI94]. The transfer behavior of the machining center was modeled as a second-order system with time delay [ADAM16b], which was a common standard, as e.g. in [CHEN98; FUSS91; ALTI94]. ADAMS [ADAM16a] predicted the future active force based on an *a priori* simulation of the engagement conditions and the current feedrate. The feedback loop of the force was closed by a repeated identification of a force model every 80 ms (indicated by the open gate in Fig. 2.5), while the MPFC controlled the feed velocity at a rate of $T_s = 20$ ms [ADAM16a, pp. 107–109]. The identification let the system consider tool wear and material uncertainties implicitly. However, the discrete nature of the identification provoked instabilities, which was why an exponential smoothing was applied to the identified model parameters introducing a delay to the feedback loop. Details on model identification in milling discusses the next section (Section 2.2).

Using the inverse of the discretely identified force model, the desired force $F_{a,des}$ was translated to a reference of the feed velocity v_f ("reference generator"). The MPFC worked in time domain with the model of the machine tool while the reference was provided per position along the tool path. Due to the formulation of the optimization problem, undesirable oscillating behavior was observed during abrupt deceleration. With a decreasing velocity, the reason for this deceleration (e.g. a change in depth of cut) could fall out of the (temporal) prediction horizon, causing the controller to accelerate again resulting in an oscillating velocity. To avoid this undesired oscillation, the reference was processed by a double filter (forward and reverse filtering) leading to a suboptimal – in special cases even unfeasible – trajectory [SCHW17]. This issue was solved by formulating the optimization problem in position domain [SCHW17; STEM20].

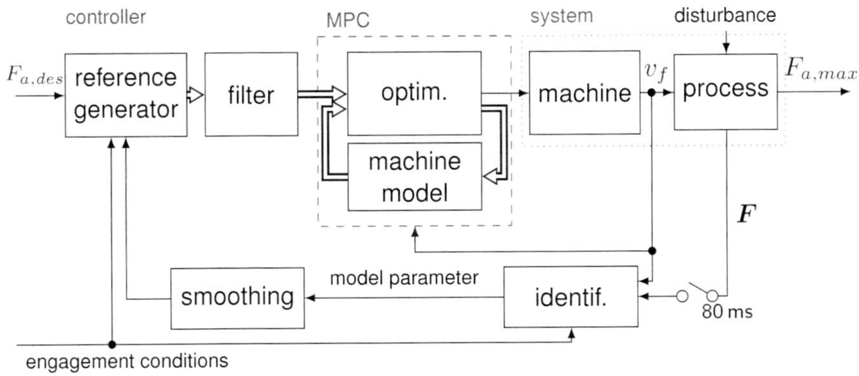

Figure 2.5: *Sketch of the model predictive force controller (MPFC) in milling by [ADAM16a]*

2.2 Process model identification

In its beginnings, the research on metal cutting was characterized by geometrical considerations on the chip and the tool engagement, e.g. [FRIE07; MART41]. Pioneering work on *"the art of cutting metals"* was conducted by TAYLOR [TAYL07]. With the publication in 1907, his work initiated a change in how to handle cutting technology and provided the concentrated findings of 26 years of laborious experiments. Although TAYLOR observed that *"the thickness of the chip or layer of metal removed by the tool has a much greater effect upon the cutting speed than any other element"* [TAYL07, p. 10] in turning, he did not yet found a functional link between cutting speed and the pressure on the tool [TAYL07, p. 119].

A vast variety of models has been proposed since then. The different modeling approaches can be grouped into the following types:

- empirical (black box),
- analytical (white box – only physical phenomena),
- shear plane model (gray box – related to the cross section of the chip),
- mechanistic (gray box – related to the cross section of the chip),
  - lumped parameter model,
  - dual mechanism model.

Reviewing the literature suggests that predominantly shear-based and mechanistic models were used to describe the force in cutting technology. Purely empirical black box models, e.g. regression or NNs, require enormous amounts of data and lack extrapolation capabilities. The other extreme, white box models that are derived only from physical constants, are nowhere near reality yet. Shear plane and mechanistic models integrate knowledge in terms of fundamental relations between the force and the chip to the model structure but rely on data-driven tuning.

2.2.1 Force model: shear-based models

A shear plane is an abstraction of the two-dimensional plane connecting chip and workpiece. This zone of transition, where the workpiece becomes a chip, is characterized by high stresses and can be observed only with a high-speed camera. The concept was further developed towards shear zones, regions around the shear planes, in which the material undergoes drastic deformation. Fig. 2.6 (b) depicts the primary, secondary, and tertiary shear zones. The shear plane occurs in the primary shear zone. Many relationships were established to determine the hypothetical shear angle Φ. It was defined in orthogonal cutting between the feed direction and the shear plane. All related the undeformed chip thickness h to its deformed value h_{def} as an indication for the occurring stresses. For example, FU ET AL. [FU84] estimated the shear angle as:

$$\Phi = \arctan\left(\frac{h\,\cos(\gamma_0)/h_{def}}{1 - h\,\sin(\gamma_0)/h_{def}}\right), \tag{2.6}$$

where γ_0 is the rake angle, Fig. 2.6 (a). Even nowadays new models are proposed, see [MATS17]. Since accurately measuring the deformed chip thickness h_{def} is almost impossible, *"shear-plane theories aim at only mean conditions at the shear and tool-chip contact planes"* [LUTT98, p. 597]. LUTTERVELT ET AL. [LUTT98] pointed out that shear plane models were popular in the American production technology while in Europe specific cutting force models evolved as standard. Since the force could be measured even in complex cutting conditions, whereas measuring the shear angle is restricted to a few conditions only, force models prevailed. Almost all are in a "mechanistic" form, relating the force to geometric values of the cross section of the chip.

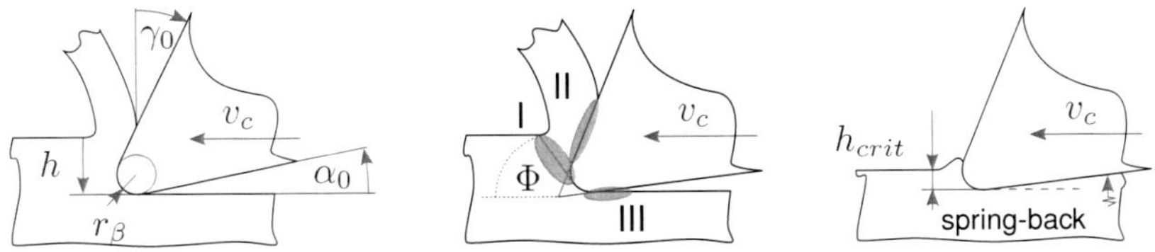

(a) geometric parameters (b) shear zones (c) ploughing

Figure 2.6: *Geometric parameters on the cutting edge*

2.2.2 Force model: mechanistic models

Mechanistic force models combine a physically motivated basis function with empirical coefficients accounting primarily for material properties and for the tool [EHMA97]. *"It combines the efficiency of analytical approaches with the accuracy of empirically calibrated models"* [WIMM18, p. 704]. The physical base is the geometry of the undeformed chip. In cylindrical milling, that is the width of undeformed chip b and the undeformed chip thickness h. The model function can be divided into two approaches: lumped parameter models and dual mechanism models.

Lumped parameter models merge the effect of shearing on the rake face and friction on the cutting edge (ploughing) in a single coefficient. Dual mechanism models, on the other hand, use two separate coefficients for these effects [ENDR95]. They are in general more accurate over a broader range of the undeformed chip thickness h [WANG04] but the ease of calculation and parameterization ("identification") of lumped parameter models were an important advantage in the beginnings of the last century. As they have become obsolete, the following refers to the dual mechanism mechanistic models simply as "mechanistic force model" as it is done in the literature nowadays.

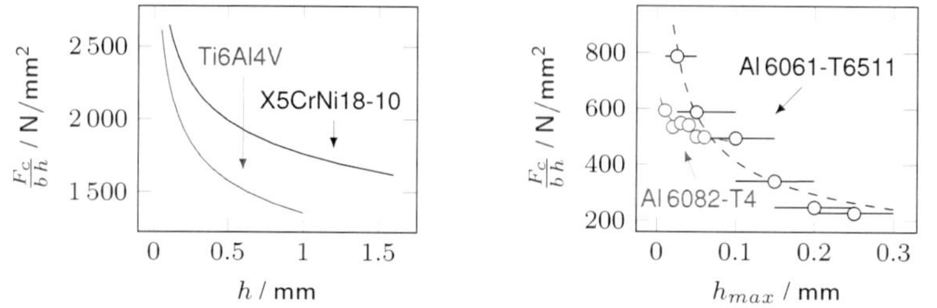

(a) *Specific cutting force according to the* KIENZLE*-model (determined in turning, [KÖNI82, pp. 58, 95])*

(b) *Specific cutting force according to the* ALTINTAS*-model (determined in milling, Al 6061-T6511 [RUBE16a, p. 318]; Al 6082-T4 [CAMP12, p. 566])*

Figure 2.7: *Specific cutting force modeled by the mechanistic force models according to* KIENZLE *(a) and* ALTINTAS *(b) over the undeformed chip thickness h (circles mark data points from literature, solid lines denote functions from literature, dashed line indicate a least squares fit of an exponential function)*

Prominent examples are the exponential force model according to KIENZLE [KIEN52; VICT56; KIEN57] and the linear approach according to ALTINTAS & LEE [ALTI96]:

$$F_{i,\text{KIENZLE}} = k_i\, b\, h^{1-m_i}, \qquad \text{where} \quad i \in \{t, r, p\}, \tag{2.7}$$

$$F_{i,\text{ALTINTAS}} = b\,(K_{i,e} + K_{i,c}\, h)\,, \qquad \text{where} \quad i \in \{t, r, p\}. \tag{2.8}$$

The parameters k_i and m_i, or $K_{i,e}$ and $K_{i,c}$ respectively, represent material-specific coefficients. They relate force to the width b and thickness h of the undeformed chip. The same functional relationship for the force F_i is assumed for the tangential force (index t), radial force (index r), and the passive force (index p). At large undeformed chip thicknesses, both models assume an almost linear increase of the force. Fig. 2.7 illustrates the specific cutting force[1] modeled with both functions indicating that the KIENZLE-model tends towards a constant value for large undeformed chip thicknesses while the ALTINTAS-model assumes an almost constant relationship at all undeformed chip thicknesses.

Mechanistic models do not claim general validity but are restricted to an admissibility range of the undeformed chip thickness h. Often, the admissibility range of linear models is extended by assuming the coefficients to be them-

[1] In turning the "cutting force" is equivalent to the feed force, while in milling, it is a synonym for the tangential force F_t.

selves again a function of the undeformed chip thickness [ZHAN17a]: e.g. a linear [GROS17], an exponential [WAN07a; WAN09b; GUO17; ZHAN18a], or a polynomial relation [WEI18; WANG18]. This second-level-modeling turns the originally linear model into a nonlinear – often exponential – model with static coefficients. Fig. 2.7b justifies this idea by illustrating the works of CAMPATELLI & SCIPPA [CAMP12] and RUBEO & SCHMITZ [RUBE16b], where the identified coefficients lay on an exponential curve. The solid lines are extrapolations of the ALTINTAS-models to the next/previous data point, on which new coefficients were identified. CAMPATELLI & SCIPPA [CAMP12] found the functional relationship of the KIENZLE-model through a study on how the feed per tooth f_t and the cutting velocity v_c affect the coefficients of the ALTINTAS-model. GHORBANI & MOETAKEF-IMANI [GHOR15] trained a NN to model the change in the specific cutting force coefficients of the ALTINTAS-model in face milling through the feed per tooth f_t, cutting speed v_c, and depth of cut a_p. Essentially, this approach joined the determination of the undeformed chip parameters and the extension of the admissible range of mechanistic models. In particular, researchers observed that the cutting coefficients $K_{i,c}$ varied with the undeformed chip thickness h [CAMP12; RUBE16b], cf. Fig. 2.7b. The trend was less obvious in the passive force F_p [CAMP12].

Studies suggested that the coefficients depended on the diameter of the tool [TSAI16], which became obvious when examining end mills with varying radius (i.e. ball or bull nose end mills) [TUKO11; WANG18; YAO18b]. GAO ET AL. [GAO13] explained this dependency with the varying cutting speed along the ball section modeling it with a cubic polynomial. Other works built this relation indirectly by relating the coefficients to the height of the tool and thereby to the arc of the ball end mill, e.g. [WANG18]. For a brief general overview of force models for ball end mills see [KURT11, p. 1051].
At high cutting velocities v_c, the temperature increases just in front of the rounded cutting edge radius r_β reducing the deformation strength so that the chip formation is facilitated [XU96, p. 49]. This thermal softening manifests itself in a downward trend in the coefficients at increasing spindle speed: RUBEO & SCHMITZ [RUBE16b] and GROSSI [GROS17] observed this phenomena by a nonlinear variation of the cutting coefficients $K_{i,c}$. The coefficients decreased with increasing spindle speed up to a critical value. From then, the observations differed: RUBEO & SCHMITZ [RUBE16b] found an upward trend while GROSSI [GROS17] observed the coefficients to be constant. Such effects may only become apparent if a large range of cutting speed is considered [WIMM18]. The nonlinear behavior at low cutting speeds was already noted at the development of the KIENZLE-model, where VICTOR [VICT56, p. 92] held the linear approximation justifiable for cutting speeds higher than $v_c > 1.25\,\mathrm{m/s}$ for solid carbide tools. Originally, the dependency on the cutting

speed was neglected because only limited ranges were relevant at the ascent of carbide tools [KIEN57].

Integrating the cutting velocity v_c, the UNIVERSITY OF ILLINOIS developed a mechanistic force model assuming an exponential relationship to the undeformed chip thickness h including also the rake angle γ_0 [KAPO98; ZHU99]:

$$F_i = b\, K_{i,m}\, h, \qquad \text{with} \quad i \in \{t, r, p\} \quad \text{and}$$
$$\ln(K_{i,m}) = a_{i,0} + a_{i,1}\, \ln(h) + a_{i,2}\, \ln(v_c) + a_{i,3}\, \gamma_0 + a_{i,4}\, \ln(v_c)\, \ln(h), \tag{2.9}$$

in which the five coefficients $a_0 - a_4$ must be identified. Neglecting the influence of the cutting velocity v_c and the rake angle γ_0, Eq. (2.9) collapses to the KIENZLE-model.

Tab. 2.2 presents an extraction of the works on the linear and on exponential models. Many of these works describe how to identify a certain experimental behavior rather than examining fundamental relationships. Therefore, the table already condenses the state of the art of force model identification, Section 2.2.5.

Table 2.2: *Overview of the literature on the usage of mechanistic force models in milling*

Model type	References
ALTINTAS-model ( ): $f(h)$, $\star$: $f(z)$	[WANG02] [WANG03a] [WANG04] [GRAD04] [RIVI09] [GONZ10] [TUKO11]* [DHUP12]* ([GAO13]) [KARA14] [ADEM15] [RUBE16a] [ZHAN16] [GROS17] [WANG18]* [YAO18a] [YAO18b]* [LIU18]
KIENZLE-model, ( ): diff. exp. model	[FU84] ([SHIN97]) ([JAYA01]) ([WAN07b]) ([WAN07a]) ([WAN08]) [WANG13] [ADEM15] [ADAM16a] ([ZHAN17b]) [SCHW19a]

radial deviation models: __ KLINE & DEVOR, ( ) models only eccentricity

Besides the two well-established mechanistic models, there is an ongoing struggle to build new models focusing on special aspects, e.g. [ENDR95; WAN09a; AUER15]. ENDRES ET AL. [ENDR95] proposed to relate the force to the interference volume (amount of material that is displaced through ploughing) rather than to the undeformed chip parameters. Calculating this interference volume proved to be impractical. AUERBACH [AUER15; AUER18] went a different way deriving a model based on the macro machining variables depth of cut a_p and width of cut a_e to facilitate the application for the process design in the industry. Other models concentrate on the radial deviation of the cutter [WAN09a; MATS17]. The models were extended to account for tool wear or chip breakage control features, i.e. primary the geometry of the rake face [ZHU99]. In general, mechanistic models, which explicitly considered the undeformed chip thickness h, outperformed models that do not consider the undeformed chip thickness [WAN09a]. The physical interpretability coupled with data-driven flexibility explains the tremendous success of the very simple functional relationships of KIENZLE and ALTINTAS & LEE.

CABRAL [CABR16] picked up the idea of modeling the influence of the microgeometry of the tool such as the rake angle γ_0 building a model with ten parameters $a_{i,0} - a_{i,9}$:

$$\frac{F_i}{b} = a_{i,0} + a_{i,1}\, r_\beta + a_{i,2}\, {r_\beta}^2 + a_{i,3}\, v_c + a_{i,4}\, {v_c}^2 + a_{i,5}\, \alpha_0 + a_{i,6}\, \gamma_0 + h\left(a_{i,7} + a_{i,8}\, v_c + \frac{a_{i,9}}{r_\beta}\right) \quad \text{with} \quad i \in \{t, r, p\}. \tag{2.10}$$

If the rounded cutting edge radius r_β is large, the effective rake angle γ_0 decreases, which in turn may lead to work hardening effects via higher deformation rates [CABR16, p. 54]. The tool orthogonal clearance α_0 affects the heat in the cutting zone.
Since the amount of ploughing depends on the size of the cutting edge rounding r_β [XU96], it is reasonable to integrate this rounding – especially when it comes to small undeformed chip thicknesses as in finish milling in general and ball end milling in particular. Nonetheless, apart from individual exceptions (e.g. [CABR16, pp. 63–65]), these effects are seldom modeled explicitly. Therefore, the model coefficients are linked to both, the material and the tool.

Hence the idea of mechanistic models originates from turning, the models are defined for the ideal cutting edge. More complex geometries, such as helical cutting edges, are sliced into stacked disk elements reminding of a spiral staircase, Fig. 2.8. Each element resembles an ideal cutting edge performing an orthogonal cutting operation. While early works formulated continuous functions for integration (e.g. [ALTI96]), they were quickly replaced by the sum over finite, discrete elements quickly replaced mathematical correctness for calculation speed.

Having a discretized model of the tool, the tool deflection can be modeled by shifting the disk elements relative to their position along the height of the tool [SUTH86]. Considering the tilt becomes necessary for long tools – for short tools, no improvement over a rigid model is observed. Originally conducted as

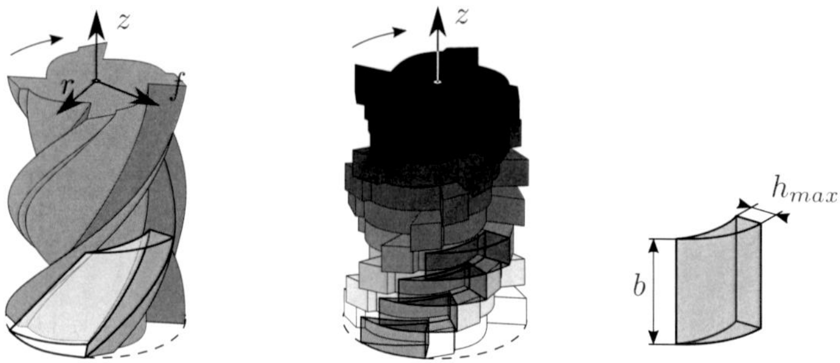

Figure 2.8: *Sliced tool with helix angle; 3D geometry of a single chip*

a trick to account for the helix angle, it pays off when the cutter experiences complex engagement conditions, e.g. peripheral face milling [ZHAN17a] or helical milling [WANG12; LI17]. The discretized tool allowed to establish different sets of model coefficients for different slices. For example, for the bottom slice, where face milling dominated, and for the upper slices, which performed peripheral milling. Other works assumed a low depth of cut a_p neglecting the helix angle, e.g. [WAN08; WANG18; YAO18a]. YAO ET AL. [YAO18b] found that peripheral milling dominates the process at a higher depth of cut $a_p > 0.6$ mm allowing to assume the same coefficients for all disk elements, including the bottom slice.

All models are only defined in the presence of a defined cut, i.e. for a defined chip. However, the cyclic intermittent nature of the milling process produces comma-like chips, where ploughing (Fig. 2.6 (c)) inevitably occurs. Undeformed chip thickness close to or smaller than the rounded cutting edge radius r_β is prone to this effect [WALD99]. WALDORF ET AL. obtained good results with modeling a stable build-up in front of the cutting edge instead of a single separation point as it was borrowed from the theory of fluid mechanics. The clearance face performs a deformation generating a friction force. The magnitude and the direction of the ploughing force vector change with increasing rounded cutting edge radius r_β [CABR16, p. 54]. Even with a perfectly sharp cutting edge, ploughing occurs due to the elastic spring-back of the material.

2.2.3 Cutting with a defined edge: undeformed chip parameters

A variety of approaches have been proposed to determine the macroscopic engagement conditions between tool and workpiece. For comparably simple tool-workpiece engagements, the geometric intersection can be described with analytic functions. Such a description has been made for turning, face milling, or flat end milling but also for five-axis machining, e.g. [WEI18]. If it comes to complex interactions as in ball end milling, simulations are used. They are either constructed from volumes via constructive solid geometry (CSG), polygons (e.g. [KAPO98]), or as virtual objects built from a finite number of elements (FEM) [WITT19]. Such elements can be directional (DEXEL, depth, e.g. [TUKO11]) or volumetric (VOXEL), e.g. [KARU09; MA19; WITT19].

Mechanistic force models rely on the microscopic perspective of this engagement, i.e. the undeformed chip geometry. The undeformed chip parameters are derived by examining the kinematics of the milling process merging the movement with the macroscopic engagement conditions along the tool path. In milling, every point of a cutting edge performs a trochoidal movement due to the tool rotation in conjunction with a superposed translational movement (feed), Fig. 2.9. Describing this movement, the undeformed chip thickness h is a function of the tool radius R, the feed per tooth f_t, and the angular position of the tooth φ_t:

$$h(\varphi_t) = f_t\,\sin(\varphi_t) + R - \underbrace{\sqrt{R^2 - \underbrace{{f_t}^2}_{(\ll R)^2}\;\underbrace{\sin^2(\varphi_t)}_{\leq 1}}}_{\approx R}\,. \tag{2.11}$$

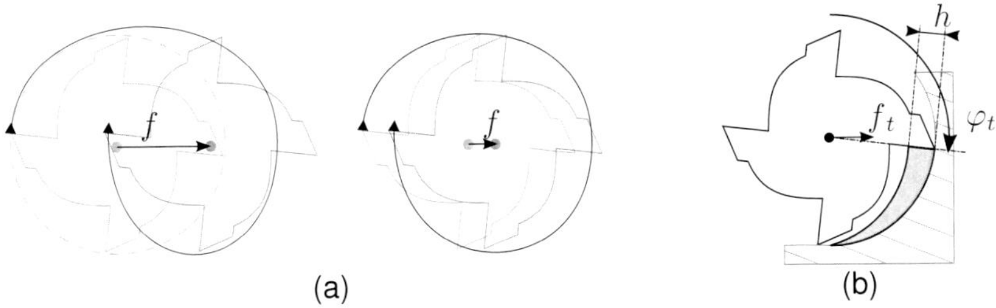

Figure 2.9: *Trochoidal path approximation (a); undeformed chip thickness (b)*

Assuming the feed per tooth f_t to be much smaller than the radius of the tool $f_t \ll R$, Eq. (2.11) can be simplified to a circular movement:

$$h = f_t\,\sin(\varphi_t)\,. \tag{2.12}$$

A different viewpoint is that the cutting velocity v_c must be much larger than the feed velocity v_f:

$$\frac{v_f}{v_c} = \frac{f_t\,N_t\,n}{2\pi\,R\,n}, \tag{2.13}$$

with n being the tool rotation and N_t the number of teeth.

The argumentation also holds in the presence of radial deviation (Section 2.2.6) as it is usually much smaller than the radius R of the tool. MARTELLOTTI [MART41, p. 693] emphasized that the minor error introduced by this approximation *"does not justify the use of the more complicated expression,"* although he examined the exact description of the deformed chip.

The reasoning is intuitive for cylindrical end mills as in face and peripheral milling but may reach its limit in ball end milling. As the radius with regard to the rotational axes tends towards zero, the square root term in Eq. (2.11) becomes significant depending on the feed per tooth. Nevertheless, usually there is no cutting at the apex of the ball.

Since this work discusses peripheral face milling, the discussion of the engagement conditions is limited to the ones of cylindrical end mills. The tool produces classical comma chips, which are fully described by the curve of the undeformed chip thickness $h(\varphi_t)$ and the width of undeformed chip b. The latter is constant for a certain chip. It depends on the orientation of the rotation axes (major axes of the tool) with regard to the surface of the workpiece. If they are oriented perpendicular to each other, the width of undeformed chip coincides with the depth of cut $b = a_p$ as is usually the case in three-axis milling.

2.2.4 Chip size effect

The coefficients of mechanistic force models tend to increase dramatically at small chip thicknesses h limiting their validity to certain regions only, cf. Fig. 2.7. This effect is called the "size effect" [WAN07a; DOTC08]. It is the manifestation of ploughing and becomes in particular apparent in the case of milling with round inserts, where it may occur in both undeformed chip parameters h and b [GHOR15]. The phenomenon has not been widely examined because the size effect occurs symmetrically in slot milling [YAO13]. When the tool enters $\varphi_{in} = 0°$ and exits $\varphi_{ex} = 180°$ with opposite directions of action, the phenomena cancel each other out. This zero-sum game may be one of the reasons why most works on model identification experiments were performed in slot milling.

Nevertheless, especially linear force models suffer from the size effect [WAN07a; DOTC08; YAO18b; YAO18a] as also could be seen in Fig. 2.7b. Exponential models tend to better represent small undeformed chip thicknesses than linear models [WAN09a; GHOR15]. YAO ET AL. [YAO18b] found their identification method to be not suited for depth of cut smaller than $a_p < 0.6\,\text{mm}$

and concluded that the linear ALTINTAS-model cannot represent these regions. WANG ET AL. [WANG12] experienced the size effect in TiAl6V4 already at $a_p \leq 0.2\,\text{mm}$. They suggested correcting the model coefficients of the ALTINTAS-model in a preprogrammed manner: decrease the coefficients by $30\,\%$ when the passive force F_p doubles in helical plunge milling. Tab. 2.3 summarizes the different limits, materials, and tool properties. Since the root cause of the chip size effect, ploughing, depends on the size of the cutting edge roundness [XU96], a threshold for the minimum undeformed chip thickness can be formulated as a function of it. CABRAL [CABR16, p. 62] suggested $h_{TH} = 0.3\ r_\beta$ for hot-work steel (X38CrMoV5-1). The ideal rounding of a cutting edge is specific to the material being machined [BERG18]. This individuality makes the size effect material-dependent in two ways: from the machinability of the workpiece itself and from the cutting edge rounding, which is again material-dependent. This double dependency makes it difficult to find a universal undeformed chip thickness, below which mechanistic force models are invalid due to the chip size effect.

Table 2.3: *Overview of the different limits up to which mechanistic force models are defined with regard to the size effect*

Reference	h_{TH} mm	Tool D mm	N_t -	β °	Material	Type	Workpiece material
[WAN08]	0.01	16	3	30	carb.	flat	Al 2618
[GONZ10]	0.06	16	1	0	carb.	flat	AISI 1045
[WANG12]	0.2	7	4	12	-	flat	Ti6Al4V
[YAO18b]	0.6	10	1	30	carb.*	bull	GH909

carb.: cemented carbide; *: with coating

2.2.5 Identification methods for mechanistic force models

The major challenge of all proposed models is to identify their parameters [KRÜG13]. The great majority of the works on how to identify the coefficients of mechanistic force models has been conducted for linear models, namely for the force model of ALTINTAS (Eq. (2.8)), Tab. 2.4. Only a few articles claimed to use a lumped parameter formulation [SHIN97; WAN07a; WAN07a; WAN08]. Of which all found an exponential dependency of the coefficients to the undeformed chip thickness $K_i(x) \propto h^x$, turning them into nonlinear models. Nonlinear dual mechanism models were identified too. Some works used an exponential basis function similar to the KIENZLE-model (Eq. (2.7)), e.g. [ZHU99; JAYA01; DOTC08; WANG13; ADEM15; ZHAN17b; ZHAN18a], or identical to it, as in [FU84; PERE13; ADAM16a; SCHW19a].

It is generally assumed that the coefficients depend only on material characteristics and the macrogeometry of the tool [KÖNI82; WANG14]. Nevertheless, the accuracy of mechanistic models depends on the range of the undeformed chip thickness. Though originally designed for *"cutting processes with constant chip thickness"*[2] [VICT56, p. 145], KÖNIG ET AL. [KÖNI82] acknowledged a small influence of the chip thickness for the KIENZLE-model by providing slightly different coefficients for different ranges of the undeformed chip thickness.

The parameter identification can be distinguished between the method of

- average forces and of
- instantaneous undeformed chip thickness.

Method of average forces

The method is inspired by turning, which exhibits a static cutting force. For different cross sections of the chip, a functional relationship is determined through linear regression [KÖNI82]. To reproduce these conditions in the cyclically intermittent cut of the milling process, the force is averaged per revolution of the tool. The arithmetic mean corresponds to the static component of a signal. To describe the functional relationship $F = f(h)$, several experiments have to be conducted at different feedrates $h \sim f_t$, Eq. (2.12). This inefficient usage of data is the most criticized drawback of this identification method, e.g. by [SHIN97; JAYA01; GROS17; ZHAN17b; SCHW19a]. However, the method of average forces is invariant to the radial deviation of the tool and to unevenly spaced cutting edges [FU84]. The lack of sensitivity can also turn into a disadvantage: *"If the tool runout [radial deviation] cannot be neglected, the average chip thickness ... is no longer valid"* [WIMM18, p. 713].

The coefficients had been determined in turning [KÖNI82] and transferred to milling. Dedicated milling coefficients usually were obtained from full immersion slot milling as it best produced signals for the method of average forces [ADEM15]. The models were identified by linear regression but KARANDIKAR ET AL. [KARA14] applied BAYES's theory to the identification problem. Although they claimed higher robustness compared to a traditional least squares fit, the method was not further pursued in the field of force model identification in cutting technology.

[2]German: *"... Zerspanverfahren mit gleichbleibender Spandicke: Drehen, Hobeln und Bohren"* [VICT56, p. 145]

A remarkable approach was shown by AGGARWAL ET AL. [AGGA13] determining the tangential cutting coefficients of the ALTINTAS-model from the current of the spindle motor with the method of average forces. The coefficients differ from coefficients identified on measurements of a rotating cutting dynamometer (RCD) by $2\,\%$ for the cutting coefficient $K_{t,c}$ and by $16\,\%$ for the cutting coefficient $K_{t,e}$.

Method of instantaneous undeformed chip thickness

Essentially, the method of instantaneous undeformed chip thickness is a curve fit between measurements and a simulated force signal based on the present engagement conditions. The method was already described in 1997 by SHIN & WATERS [SHIN97], but only became much later attractive.

Curve fits for model identification have been dominated by global optimization algorithms such as evolutionary algorithms [GHOR15; GROS17; CHEN18] or particle swarm algorithms [ZHAN17b]. This preference was independent of whether it was the linear ALTINTAS-model [GROS17] or the exponential KIENZLE-model [ZHAN17b]. Nevertheless, local optimization algorithms have a significant advantage in computation with no loss in accuracy [SCHW19a]. They can be real-time capable and – if one uses constraint algorithms – can ensure technological plausibility. FREIBURG ET AL. [FREI15] even reported a lack of convergence if unconstrained optimization was used. GONZALO ET AL. [GONZ10] built their own constraints by limiting divergence from the initial guess.

Comparing the methods of average forces and instantaneous undeformed chip thickness

ADEM ET AL. [ADEM15] compared model structures and identification methods: the linear ALTINTAS-model to the exponential KIENZLE-model and the method of average forces to a least squares curve fit (method of instantaneous undeformed chip thickness). They concluded that the nonlinear force model was generally more accurate than the linear model. A finding supported by more recent works [WEI18; WIMM18]. They further found that the optimization-based curve fit resulted in more accurate models than the average forces approach [ADEM15; WIMM18]. This advantage could not be confirmed by SHIN & WATERS [SHIN97], who obtained almost equal coefficients for both methods for the KIENZLE-model. However, for the linear ALTINTAS-model, RUBEO & SCHMITZ [RUBE16b] found much higher coefficients for the cutting force component $K_{i,c}$ with the method of instantaneous undeformed chip thickness (up to twice the value). WIMMER ET AL. [WIMM18] claimed that the method of average forces was more robust to oscillations in the measurements and, thus, should be used for finish milling (with small chip sizes). For a systematic comparison of both identification methods, GONZALO ET AL. [GONZ10] defined an ideal reference for the linear ALTINTAS-model. They identified the coefficients in turning defining these values as the "single point of truth". Comparing them to coefficients obtained in milling, they concluded that the method of instantaneous forces better matched the "true" coefficients and thus had better physical credibility. By using a single tooth cutter, they avoided the problem of radial deviation taking away the advantage of the method of average forces.

An interesting combination of both methods performed WAN ET AL. [WAN07a] and GUO ET AL. [GUO18]. They averaged the force measurements at the same instantaneous chip thickness of all teeth. Assuming a constant tooth spacing, this approach can be exemplified as:

$$\overline{F}(\varphi) = \frac{1}{N_t} \sum_{j=0}^{N_t-1} F\left[h\left(\varphi + \frac{360^\circ}{N_t} j\right)\right], \tag{2.14}$$

where N_t is the number of teeth and φ is the cutter rotation angle. The line above the symbol $\overline{F}$ indicates the average of the force. This method combines the insensitivity of the method of average forces to the radial deviation with the high utilization rate of the force signal of the method of instantaneous undeformed chip thickness.

Besides these two major identification methods, a few exceptional approaches have been proposed. While in the early 2000s some works identified the force models via frequency analysis e.g. [SEET99; JAYA01; WANG03b; WANG04] and WANG & CHANG [WANG04] built a force model directly from the FOURIER series. An entirely different approach presented DOTCHEVA ET AL. [DOTC08] and WANG ET AL. [WANG13] using the surface error to determine the coefficients of an exponential force model via the tool deflection and the force acting on it.

Tab. 2.4 summarizes the described state of the art. It gives an overview of the different identification methods and marks the model types. Recent works seem to favor the method of instantaneous undeformed chip thickness. TUKORA & SZALAY [TUKO11] mentioned the idea of an online identification of a force model as a concept. ADAMS [ADAM16a, 107 ff] identified the model coefficients online within three cutter revolutions, which translated to 80 ms allowed execution time or 1 mm tool path. At each of these discrete identifications, a new, independent model was identified. Changing the model coefficients discretely provoked instabilities through oscillating feed commands (manipulated variable). Therefore, he applied recursive smoothing, essentially leading to a delay of the new model parameters of 400 ms. This delay in the feedback of the force control loop represented 20 time steps of the controller. SCHWENZER ET AL. [SCHW19a] showed that there is only a small potential to further improve the computation for curve fits. STEMMLER [STEM20] applied this improved version of the discrete identification to force control of the milling process. He confirmed the necessity of smoothing the identification for the proper performance of the controller. ALTINTAS & ASLAN [ALTI17] picked this idea up using a recursive least squares (RLS) fit.

Table 2.4: *Overview of the literature on the identification of mechanistic force models*

Identification method	References
Average forces	[KIEN52]‡ [KÖNI82]‡ [WANG02]⋆ [GRAD04]⋆ [GONZ10]⋆ [WANG13]‡ [GAO13](⋆) [WANG14](⋆) [KARA14]⋆ [ADEM15]⋆‡ [TSAI16]⋆ [RUBE16a]⋆ [WANG18]⋆ [LIU18]⋆
Instantaneous undeformed chip thickness ( ): averaged	([WAN07a](‡)) [WAN08](‡) [DOTC08](‡) [RIVI09]⋆ [GONZ10]⋆ [TUKO11]⋆ [DHUP12]⋆ [FREI15]‡ [ADEM15]⋆‡ [ADAM16a]‡ [ZHAN16]⋆ [RUBE16a] ⋆ [GHOR15]⋆ [ZHAN17b](‡) [GROS17]⋆ ([GUO18](‡)) [SCHW19a]‡
FFT-based	[JAYA01](‡) [WANG03b]⋆ [WANG04](‡)

⋆: ALTINTAS-model, (⋆): linear force model,
‡: KIENZLE-model, (‡): nonlinear force model
radial deviation models: ____ KLINE & DEVOR; _ _ _ _ other

2.2.6 Radial deviation model

Radial deviation[3] of a tool produces a dissimilar feed per tooth and, thus, a periodically varying chip load. It affects tool life, surface finish, and even the stability of the machining process [SCHM07].

The progression of tool wear may lead to contradictory results: As the cutting edge recesses more teeth become eventually involved, resulting in a more homogeneous load as it was originally intended [BAUM19]; but if the protruding

[3]The term "runout" is a widespread synonym for radial deviation. This work uses the terms defined in the CIRP Dictionary of Production Engineering [INTE97] for the sake of consistency.

cutting edge removes fatally before the others can unload it, a domino effect is initiated. Knowing the radial offset enables to individualize stability lobe diagrams for a distinct tool to prevent chatter [SCHM07].

The most common model to describe the radial deviation of milling tools was postulated by KLINE & DEVOR [KLIN83]. They modeled it as a parallel offset and an additional tilt of the axis of rotation, Fig. 2.10. The error manifests itself in a different local radius for each tooth t of all N_t teeth:

$$R_t(z) = R + \underbrace{\rho \cos\left[\delta\lambda(z) - \lambda + \frac{2\pi\,(t-1)}{N_t}\right]}_{\text{offset}} + \underbrace{(L-z)\,\tan(\tau)\,\cos\left[\delta\lambda(z) - \lambda - \psi + \frac{2\pi\,(t-1)}{N_t}\right]}_{\text{tilt}}, \tag{2.15}$$

where R is the nominal radius of the cutter, ρ the eccentricity or radial offset of the axis of rotation, and λ denotes the corresponding angle of direction (polar coordinates). The direction of the offset varies along with the height of the tool (and hereby with the current disk element) due to the helix angle β:

$$\delta\lambda(z) = \tan(\beta)\, z. \tag{2.16}$$

The parameters τ and ψ in Eq. (2.15) describe the tilt angle and the angle of direction of the tilt respectively. The tilt depends on the unsupported length L of the tool, which is measured between the tip of the tool holder and the tip of the tool. The effective radius $R_t(z)$ varies with the height z of the tool due to the tilt and is unique for every tooth t. The equations in this section assumes all angles to be given in rad.

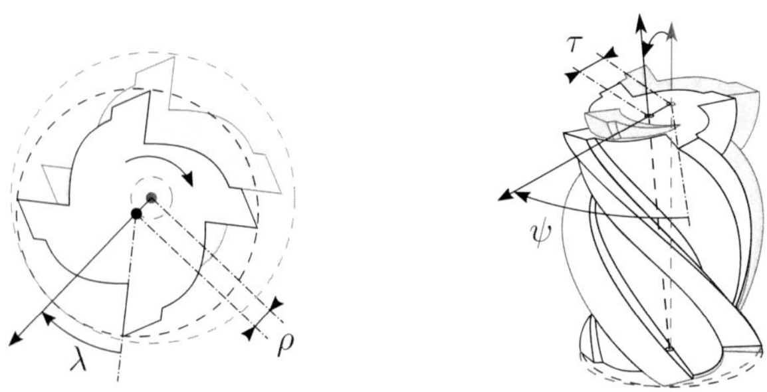

Figure 2.10: *Sketch of the radial deviation model: offset (left) and tilt (right) of the principal axis*

With a minor geometric approximation ($R^2 >> {f_t}^2 \cos^2(\beta)$ [KLIN82, p. 35]), the undeformed chip thickness $h_t(z)$ along the height z of the tooth t can be expressed as:

$$h_t(z) = \underbrace{R_t(z) - R_{t-1}(z)}_{\Delta h(z,\rho,\lambda,\tau,\psi)} + \underbrace{f_t \sin(\varphi_t)}_{h_{\text{nominal}}}. \tag{2.17}$$

The exact geometric derivation describes KLINE [KLIN82] in detail in his dissertation. To consider large radial deviations (close to the feed per tooth), he suggests adding an individual constant offset for every tooth $R_t = R + C_t$. This extended formulation is seldom used in other works.

The model is very popular due to its simplicity. Tab. 2.2 emphasizes this observation by illustrating that this model is almost exclusively used. Since the magnitude of the radial offset (eccentricity) usually dominates the radial deviation, the cutter tilt is often neglected [PERE13]. Modeling the tilt is only important for long, slender tools. The improvement when considering it for all kinds of tools is only marginal [WIMM18]. Therefore, most works follow the simplification of neglecting the tilt when describing the radial deviation, e.g. [WAN07b; ADAM16a; ZHAN17b; ZHAN16; GUO18].

PEREZ ET AL. [PERE13] further simplified the radial deviation model using the averaged undeformed chip thickness $\overline{h(z)}$. Using average values saved them from discretizing the tool but still showed good accordance to a classical "sliced" model. However, they advised adjusting the start and exit angle of engagement φ_{in}, φ_{ex} due to radial deviation, introducing higher complexity to the modeling of the process. They acknowledged that in most published papers the engagement angle is not updated. This simplification is generally accepted since the first days of this model [FU84].

2.2.7 Identification methods for radial deviation models

If considered at all, the radial deviation usually has been described only by the offset ρ and the corresponding rotation angle λ, Eq. (2.15). Even this simple model can improve the description of the force by $30\,\%$ [RIVI09]. The parameters are either measured [JAYA01; WOJC15] or identified as well [WAN09b; ZHAN16; WEI18].
One can identify the radial deviation separately from the force model, cf. [WAN07a; WAN08; WANG13; ZHAN16], or jointly with the force model, cf. [RIVI09; FREI15; RUBE16a; ADAM16a; SCHW19a]. Picking up the idea of "averaged instantaneous undeformed chip thickness" allows to separate the part of the force that is caused by the radial deviation. CHEN ET AL. [CHEN18]

proposed to use an evolution-inspired global optimization to identify the radial offset.

The radial deviation changes with the spindle speed, because the tool-holder-spindle system exhibits different dynamic characteristics at different speeds [ZHAN17a; WIMM18]. Therefore, ZHANG ET AL. [ZHAN18a] emphasized identifying the radial deviation at the same spindle speed as in the cutting process, because the rotation causes a centrifugal force on the spindle yielding to an elastic extension, which in turn affects the radial offset.

The periodic excitation of a radial deviation inspired authors to identify the radial deviation in the frequency domain, e.g. [SEET99; WANG03b]. WANG & ZHENG [WANG03b] determined the radial offset ρ from the shear component of a linear force model combining the identification of the force model and of the radial deviation model. The approach required two separate experiments at slightly different feeds to determine the radial offset through the harmonics in a fast FOURIER transformation (FFT). For identification, they expressed the force model including the effect of the radial offset as a FOURIER series. SEETHALER & YELLOWLEY [SEET99] built on the linear ALTINTAS-model and identified the radial deviation of each individual tooth. However, the frequency domain only allows to quantify the radial offset ρ but may not provide any information about the direction λ or of the tilt τ, ψ.

2.3 Summary and findings on the state of the art

The main building blocks addressed here were the control of the (active) force in milling and the identification of a mechanistic model of the milling process.

The milling process is characterized by abrupt and drastic changes in the process force due to sudden changes in the engagement of the tool. These step-like changes in the force put a challenge on the control of it. Process control is not yet common in industry – neither of the force as control variable, nor of alternative quantities such as electrical power. The low prevalence is partly due to the insufficient performance of classic control in the presence of fast-changing engagement conditions and partly due to the laborious set-up and tuning of most control systems, which requires expert knowledge. Since classical controllers only react to deviations of their controlled variable from the reference, they inevitably fail at sudden changes in the engagement conditions (Section 2.1.2). ULSOY & KOREN [ULSO89] blamed the bad performance on the lack of prior knowledge about the process. Nowadays, the *a priori* available knowledge has vastly increased through simulations, which are now an integral part of a proper process design. And still, the process has to contain a safety margin because the knowledge is just not complete – and never can be due to the inherent uncertainties of the real world.

Early, the idea arose that combining a geometric simulation and a controller could bridge this gap (Section 2.1.4). Usually, the geometric engagement was calculated offline prior to the online part of closed-loop control. One of the most successful classic controls for the milling process used an adaptive controller in conjunction with an offline simulation and an online updated force model [ALTI17]. And still, reactive control must fail at abrupt changes risking tool failure.
Methods from the field of "advanced control" suggest a promising approach to handle this inherent challenge of the milling process. Among them, the anticipating nature of MPC proves to be especially well-suited for the milling process (Section 2.1.5). It solves a potentially constrained optimization problem for a finite prediction horizon at every step using a model of the process (Section 2.1.5). In this way, a new control law is determined implicitly and tuning is allowed in the often more intuitive time domain. While at the beginning feasibility was prime [ADAM16a], latest studies addressed usability and tuning of the MPC for force control in milling [STEM20]. Others dropped simplifying assumptions regarding the machine tool model [AY19a].

Regarding desired constraints and physical limitations, ALTINTAS & MA [ALTI90] included the actuator saturation in their discontinuous machine model, enabling the adaptive controller to consider a truncated nonlinearity. STEMMLER ET AL. [STEM16] and ADAMS [ADAM16a] explicitly considered constraints in both, the manipulated variable (v_f) and the control variable (F_a), using MPC. In fact, the MPFC implemented (constrained) optimal control at each sampling point.

To model the force in cutting, so-called *mechanistic* models dominate the field (Section 2.2.2). They link physically reasonable values (the undeformed chip geometry) with data-driven coefficients. These coefficients are specific for a given tool-workpiece combination and cutting velocity. The models constitute an appreciated trade-off between generality, accuracy, and computational effort. KIENZLE [KIEN52] suggested an exponential relationship between the force and the undeformed chip thickness, while ALTINTAS & LEE [ALTI96] proposed a linear model with a constant offset. However theses linear coefficients were often themselves modeled as a function of the chip turning the model into a nonlinear model.

To determine the undeformed chip thickness accurately, the pulsating effect of the radial deviation of the tool must be considered. It is described as a radial eccentricity and an axial tilt almost exclusively with the model according to KLINE & DEVOR [KLIN83] (Section 2.2.6). The tilt is only pivotal for slender tools and thus most often neglected.

Usually, a process model was calibrated offline at straight cuts, mostly full immersion slot milling [PLET18]. Recent publications discussed online identification of process models suggesting a recursive identification [ALTI17] or repeated discrete identification based on a curve fit [ADAM16a; STEM20].

In a nutshell:

- MPC can handle the abrupt changes in the force of the milling process because its model and an accompanying simulation of the engagement enable a prediction of the future process along the tool path. It requires an online calibration of the process model (model of the force and of the radial deviation) to close its feedback loop.
- The force model according to KIENZLE [KIEN52] presents a generic representative of mechanistic force models due to its nonlinear function.
- An online identification of mechanistic force models is usually conducted with repeated curve fits (method of instantaneous undeformed chip thickness).

3 Objectives and tasks

Reviewing the state of the art emphasizes that advanced control methods are necessary to reliably control the maximum active force in milling. MPC is a promising approach for this purpose. It allows maximum productivity while guaranteeing geometric quality and process safety. Identifying the process model online takes process uncertainties into account, such as tool wear or material properties. This identification provides the feedback in the force loop of a MPFC. There, previous works on MPC for force control in milling ([ADAM16a; STEM20]) lack an immediate identification. They relie on a hybrid control similar to two cascaded control loops running at different sample times. The MPFC performes a closed-loop velocity control, for which the reference is updated at a lower rate (20 times lower).

Therefore, this work claims that:

Hypothesis

The feedback loop of model-based predictive control (MPC) for force control in milling can be closed by an instantaneous, single-step identification of the process model.

The hypothesis can be broken into individual research questions. Since a controller requires delay-free feedback, the largest and most critical task is to parameterize a model of the milling process in real-time. The calculation is performed within a defined time, e.g. the sampling time T_s of the controller, i.e. within a single control step. The single-step-requirement demands a deterministic, hard real-time calculation because it is ultimately used in a control loop.

Hence the feedback is not a measurement signal but a model, the feedback also has to be mathematically continuous. (ADAMS [ADAM16a] and STEMMLER [STEM20] achieve continuity by exponential smoothing the uncorrelated results of the curve fit trading in “delay-free” for “continuity”.) In the context of a digital system, the demand for a continuous function suggests a relation to previous values. Based on these considerations the first research question can be phrased:

Research question 1

How to recursively parameterize a mechanistic force model and a model of the radial deviation?

The result is a general method for online identification of (mechanistic) process models in milling and represents the originality of this work. The solution applies to all mechanistic force models.

While the first question addresses a single module of the larger overall system (MPC, identification, machine tool, sensors, signal processing, etc.), the effect on the other components must be examined:

Research question 2

How does the overall system have to be designed to handle the immediate feedback when closing the feedback loop?

This question focuses on what consequences arise for the entire system from a closed feedback loop, and how the individual modules must be revised to support the novel identification method. The controller is unaware of a time delay in its feedback loop induced by a curve-fit-based identification. Hence, answering this question enables for a delay-free MPFC in milling for the first time.

Eventually, a critical assessment is demanded:

Research question 3

What is the difference in the performance of the novel control of the milling process to existing approaches?

For answering this question, the MPFC with continuous feedback must be compared to the state-of-the-art approaches (i.e. the MPFC with a repeated curve fit for identification), as well as to the current state of the industry, which in most cases is still a process with constant feed. The accuracy of tracking the desired reference serves as evaluation criterion for the control; the economic benefit may be assessed by the process time.

The layout of the given thesis is as follows: in Chapter 4, a method for a recursive estimation of mechanistic process models using a particle filter is developed. The method estimates the model like a disturbance observer using historic data from previous time steps rather than building independent models for short time windows. The novel approach is studied in detail in simulations, which allows to know the ground truth. This knowledge enables to examine the convergence behavior of the identification. It is benchmarked against a conventional curve fit and against a recursive curve fit. Afterwards (Chapter 5), the developed identification is transferred to real measurements. A small excursion demonstrates that this method prepares the ground to enhance the identification through sensor fusion. Somewhat redundant information with different quality is used to obtain a more accurate estimation. Both chapters address the first research question. The second research question is discussed in Chapter 6, in which the overall control system is designed (tuning of the MPFC, fitting a machine tool model, setting up communication, and designing computational budgets). The third research question is examined in Chapter 7, evaluating the MPC system for force control in milling. The work closes with a summary and conclusion.

4 Instantaneous process model identification in milling

The identification of a force model is an integral part of MPC for force control in milling as it was proposed by ADAMS and STEMMLER [ADAM16a; STEM16; SCHW17; STEM20]. The identification closes the feedback loop of the controller. The more accurate the model, the more accurate can the MPFC control the process at the desired level. Since MPC performs a local optimization at every sampling step it benefits from a computationally lightweight process model. The state of the art (Section 2.2) showed that mechanistic force models are suited for this purpose. This work uses the KIENZLE-model, Eq. (2.7), for the force and the KLINE & DEVOR-model, Eq. (2.15), to account for the radial deviation of a tool. Combined, both models build the process model.

The classical identification of mechanistic force models requires extensive experiments (method of average forces), whereas newer approaches perform a curve fit on a single experiment (method of instantaneous undeformed chip thickness). Nevertheless, at minimum, they require the measurements of a full cutter revolution constituting the maximum resolution in time. While theoretically this problem can be overcome by a moving window approach, in practice the large computational effort of a curve fit prevents determining a new result at every time step. Previous studies showed that an optimized fit takes 80-100 ms [ADAM16a, p. 130; SCHW19a] disqualifying this approach for an online identification in less than the desired $T_{s,MPC} = 20\,\text{ms}$ of the controller (Section 6.2). Further, the approach generates independent models but continuous feedback is desired.

A recursive formulation of the curve fit may be appropriate, e.g. a recursive least squares (RLS). Preliminary works showed its lack of robustness for force model identification [SCHW19b]. Following the idea of a recursive identification leads to observer-based methods, which weigh the predictions of a model against the measured values. The methods – e.g. the KALMAN filter (KF) – require "observability." Observability demands a unique relation between internal states and the input and output of a model. *"Observability is the issue of whether the state of a dynamic system with a known model is uniquely determinable from its inputs and outputs. It is essentially a property of the given system model"* [GREW15, p. 55]. This means that there are no two states that produce the same result – with the same input given. A state space model can be tested for observability with the KALMAN criterion [KALM59], which checks the rank of a matrix built from the system matrix and the observation matrix.

The KIENZLE-model and the radial deviation model, Eq. (2.15), are nonlinear and unobservable. This applies also to their linearized versions.

The objective of this chapter is to develop a method, which updates a model of the milling process in real-time $T_{s,idify} \ll T_{s,MPC} = 20\,\text{ms}$. In preliminary studies, the ensemble KALMAN filter (EnKF) as a truly nonlinear estimator was introduced for the problem of identifying a mechanistic force model [SCHW19b; SCHW20]. It is a particle filter or "sequential MONTE CARLO method." Such particle-based methods approximate the probability density of the model parameter rather than the mathematical expectation. So instead of approximating a single set of parameters, particle filters approximate the underlying posterior distribution by many state vectors (these vectors are the "particles") with initially different random values. Targeting the probability of the states is the reason why the model does not have to be observable and distinguishes the method from the classic approaches such as a KF. The unscented KALMAN filter (UKF) [JULI97] is a similar, also particle-based, method but propagates special points forward in time – so-called "sigma points". Thus, the UKF is not based on a MONTE CARLO simulation, which requires random points. Literature disagrees on which method is better. Although the UKF is particularly elegant for model adaptation problems, the EnKF offers a slightly broader approach for inverse problems without previous knowledge on the probability distribution of the parameters.

This chapter describes the original EnKF to make use of its well-posedness. A classic (discrete) curve fit and the RLS serve as a benchmark. RLS is a standard procedure in the field of model identification and was used in [ALTI17]. Its mathematical description can be found in Appendix A.2.1.

4.1 On modeling the process force in milling

"A vast majority of past efforts deals with developing analytical models with very limited experimental validation of the models under only a few selected cutting conditions, tool geometries and work materials" [LUTT98, p. 589].

Thus, this work uses a simple, mechanistic force model and calibrates it online. Without lack of generality, this work uses the nonlinear KIENZLE-model, which already proved to be an adequate choice for an MPFC in milling [ADAM16a; STEM16]. The state of the art revealed that it is the more generic case of the well-established mechanistic force models. Moreover, a nonlinear model underlines the general applicability of this novel approach.
To build a model of the process, the force is joined with a description of the effect of the radial deviation of the tool, which is modeled in literature

predominantly by the KLINE & DEVOR-model. This section describes how the process model was implemented for this work.

To obtain the force that acts on the tool, the process model has to be mathematically integrated along the cutting edge. The integration reduces to a summation along the height of the tool if the tool is discretized as described in Section 2.2.2 and Fig. 2.8. The force of the milling process can be modeled as a function of the current position of the cutting edge φ_t (angle), which presents a measure of time:

$$F_i(\varphi_t) = k_i \, \frac{b}{N_s} \sum_{n=1}^{N_s} \phi(n) \sum_{j=1}^{N_t} \Bigg[\underbrace{f_t \, \sin \overbrace{\left[\varphi_t - (n-0.5)\, \Delta\varphi_\beta\right]}^{\varphi_{t,s}}}_{h_{\text{nominal}}(n)} - \tag{4.1}$$

$$\underbrace{2\rho \, \sin \left[\frac{\tan(\beta)}{R} \frac{b}{N_s} (n-0.5) - \lambda + \frac{\pi}{N_t} (2\,j-3) \right] \, \sin\left(\frac{\pi}{N_t}\right)}_{\Delta h_{\text{offset}}(n,j)} \Bigg]^{1-m_i} .$$

The equation uses a rewritten representation of the radial offset (Eq. (2.15)-Eq. (2.17)). With the help of the addition theorem of two arbitrary angles ϑ and ν:

$$\cos(\vartheta) - \cos(\nu) = -2 \, \sin\left(\frac{\vartheta + \nu}{2}\right) \, \sin\left(\frac{\vartheta - \nu}{2}\right),$$

the radial offset becomes

$$\Delta h(z,j) = -2\rho \, \sin \left[\frac{\tan(\beta)}{R} \, z - \lambda + \frac{\pi}{N_t} \, (2j-3) \right] \, \sin\left(\frac{\pi}{N_t}\right). \tag{4.2}$$

All angles are represented in rad for conciseness. Nonetheless, for the sake of intuition, the values will be given in degree (°) throughout this work.

Assuming equally spaced cutting edges, the position of each edge j becomes $\varphi_t = \varphi + \frac{2\pi}{N_t} \, (j-1)$. The helix angle β makes the cutting edge tilt backward from the tip. The amount of this projection $\Delta\varphi_\beta$ per slice is:

$$\Delta\varphi_\beta = \frac{b}{N_s \, R \, \tan(\pi/2 - \beta)}. \tag{4.3}$$

The model uses the center of each disk element $(n - 0.5)$ to describe the height along the tool. In peripheral end milling, the major axis of the tool is always perpendicular to the feed direction. Therefore, the undeformed chip width b is equivalent to the depth of cut: $b = a_p$. Lastly, the BOOLEAN-function $\phi(n)$ models whether a slice is engaged or not (each slice has an individual position of the cutting edges $\varphi_t(n)$):

$$\phi(n) = \begin{cases} 1 & \varphi_t(n) \geq \varphi_{in} \text{ and } \varphi_t(n) \leq \varphi_{ex} = 180°, \\ 0 & \text{otherwise.} \end{cases}$$

Peripheral end milling can be described by the following macroscopic engagement conditions:

- feed direction angle α_f,
- depth of cut a_p,
- the start angle φ_{in} of the engagement, and
- the end angle φ_{ex} of the engagement.

The feed direction angle α_f represents the angle between the feed direction and the X-axis of the machine tool. It is used for the transformation between the machine coordinate system $(\mathtt{X}, \mathtt{Y})$ and a tool-fixed coordinate system $(\mathtt{x}, \mathtt{y})$ in the feed direction. Fig. 4.1 illustrates the variables. The depth of cut a_p is perpendicular to this sketch.

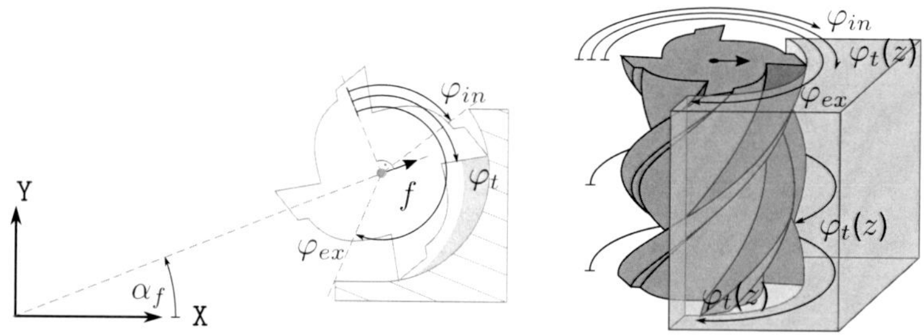

Figure 4.1: *Engagement conditions on the cutting edge*

The accuracy of the force model depends on the resolution of the tool in terms of disk elements N_s. The computation time scales linearly to it $\mathcal{O}(N_s)$. Allowing a maximum error in the resolution of the position of the cutting edge $\Delta\varphi_{t,max}$, the required number of disk elements can be estimated by the fraction of the depth of cut and the axial resolution Δz rounded down:

$$N_s = \lfloor \frac{b_{max}}{\Delta z} \rfloor \qquad \text{with} \qquad \Delta z = \frac{D \, \sin(\Delta\varphi_{t,max})}{\tan(\beta)}. \tag{4.4}$$

Where b_{max} represents the maximum width of the undeformed chip on a tool path. In peripheral face milling, this width is equivalent to the maximum depth of cut $a_{p,max}$. For example, a tool ($D = 10$ mm, $\beta = 46°$) would be approximated through $N_s = 23$ disk elements if the maximum depth of cut was $a_p = 2\,\mathrm{mm}$ and the desired angular resolution $\Delta\varphi_{t,max} = 0.5°$. These values translate to an axial resolution of $\Delta z = 0.087\,\mathrm{mm}$ – other works use a resolution of $0.02\,\mathrm{mm}$ to describe a more complex ball end mill [CABR16, p. 71]. This formulation of a process model presented in this section is used throughout this work.

4.2 Ensemble KALMAN filtering for force model identification

EnKF is a particle-based sequential optimization that offers a cheaper derivative-free approximation than traditional methods [IGLE13; CHAD19]. Finding parameters of a model based on observations is an inverse problem. The related keyword in terms of EnKF is "ensemble KALMAN inversion." Inverse problems aim at determining the inputs $\boldsymbol{u}$ of a system model $\boldsymbol{f}(\cdot)$ based on known observations (measurements) $\boldsymbol{z}$ with some optional internal states $\boldsymbol{x}$:

$$\boldsymbol{z} = \underbrace{f(\boldsymbol{u}, \boldsymbol{x})}_{\boldsymbol{y}} + \boldsymbol{\eta}. \tag{4.5}$$

The observations are subject to noise $\boldsymbol{\eta} \sim \mathcal{N}(\boldsymbol{0}, \boldsymbol{\Gamma})$ with zero mean and of covariance $\boldsymbol{\Gamma}$. The noise is assumed to have a GAUSSIAN shape – the underlying probability theory also assumes a GAUSSIAN prior and posterior probability to derive a linear analysis.

4.2.1 Excursion: The ensemble KALMAN filter

Since the method is not yet state of the art and is still completely unknown in the field of cutting technology, this section describes the underlying theory of EnKFs from an engineering perspective.

Functional principle of the ensemble KALMAN filter

Introduced by EVENSEN [EVEN94], the idea was to propagate a group of state vectors (that are the particles) forward in time and to take its mean as the best guess of this ensemble – in contrast to this, the classic KF propagates a single state vector. The propagation of this ensemble is done as a sequential optimization, see e.g. [SANZ19]. So the backbone is a MARKOV chain MONTE

CARLO simulation of the probability density of the states [EVEN03].[1] The EnKF constructs the empirical covariance of the ensemble eliminating the need to propagate the covariance in time as in the classic KF. The ensemble (of the model states) approximates the underlying probability density defining the error covariance inherently as its spread. In fact, the error of this approximation decreases proportionally to $1/\sqrt{J}$, where J is the ensemble size [EVEN03]. The covariance matrices represent how much to trust the model rather than the measurements.

Assuming an ensemble matrix of J members $\boldsymbol{x}^j_{k|k-1}$ at time step k based on the information of the previous time step $k-1$:

$$\boldsymbol{X}_{k|k-1} = \left\{\boldsymbol{x}^1_{k|k-1}, \dots, \boldsymbol{x}^J_{k|k-1}\right\}. \tag{4.6}$$

The capital $\boldsymbol{X}$ denotes the entire ensemble as a matrix of the individual state vectors $\boldsymbol{x}^j$. The potentially nonlinear model $\boldsymbol{f}(\boldsymbol{u}, \boldsymbol{x})$ predicts the output of each ensemble member before calculating the empirical covariance matrix $\overline{\boldsymbol{P}}$ (this is the prediction step of the filter):

$$\hat{\boldsymbol{X}}_{k|k-1} = \boldsymbol{f}\left(\boldsymbol{u}_k, \boldsymbol{x}^j_{k-1}\right) \qquad j \in \{1, \dots, J\}, \tag{4.7}$$

$$\overline{\boldsymbol{P}}_k = \frac{1}{J}\left(\hat{\boldsymbol{X}}_{k|k-1} - \overline{\boldsymbol{X}}_{k|k-1}\right)\left(\hat{\boldsymbol{X}}_{k|k-1} - \overline{\boldsymbol{X}}_{k|k-1}\right)^{\mathsf{T}}, \tag{4.8}$$

where lines $(\overline{\cdot})$ denote mean values and hats $(\hat{\cdot})$ predicted or estimated values. The empirical covariance matrix $\overline{\boldsymbol{P}}_k$ is calculated once per step based on the forecasted ensemble $\hat{\boldsymbol{X}}_{k|k-1}$. The predicted state vectors $\hat{\boldsymbol{x}}^j$ are corrected through observations $\boldsymbol{z}^j$ weighted by the KALMAN gain $\boldsymbol{G}_k$ (this is the update step of the filter):

$$\boldsymbol{G}_k = \overline{\boldsymbol{P}}_k \, \boldsymbol{H}^{\mathsf{T}}_k \left[\boldsymbol{H}_k \, \overline{\boldsymbol{P}}_k \, \boldsymbol{H}^{\mathsf{T}}_k + \overline{\boldsymbol{R}}_k\right]^{-1}, \tag{4.9}$$

$$\boldsymbol{x}^j_{k|k} = \hat{\boldsymbol{x}}^j_{k|k-1} + \boldsymbol{G}_k \left[\boldsymbol{z}^j_k - \boldsymbol{H}_k \, \hat{\boldsymbol{x}}^j_{k|k-1}\right], \tag{4.10}$$

where $\boldsymbol{H}$ denotes the output matrix, i.e. those states that can be measured $\boldsymbol{y}^j = \boldsymbol{H} \, \hat{\boldsymbol{x}}^j$. Note that $\boldsymbol{y}$ is the model output, while $\boldsymbol{z}$ represents noisy observations, i.e. $\boldsymbol{z} = \boldsymbol{y} + \boldsymbol{\eta}$. The observations are artificially augmented with

[1] "MARKOV chain", i.e. the next state only depends on the current state and a finite number of previous states – here, it is a 1st order MARKOV chain because the prediction depends only on the current state. "MONTE CARLO", because it uses the same model but multiple random initial values.

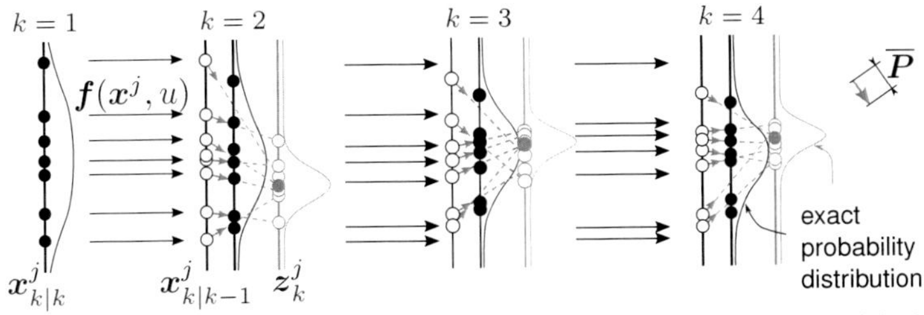

Figure 4.2: *Sketch of how the EnKF evolves its ensemble members (particles)*

zero-mean GAUSSIAN noise $\eta \sim \mathcal{N}(\mathbf{0}, \mathbf{\Gamma})$ to create individual observations for every ensemble member:

$$z^j_k = y_k + \eta^j_k, \qquad \text{with } j \in \{1, \ldots, J\}, \tag{4.11}$$

$$\overline{R}_k = \frac{1}{J}\, \eta_k\, \eta_k^{\mathsf{T}} \approx \mathbf{\Gamma}. \tag{4.12}$$

In the limit of infinite ensemble members $J \to \infty$, the empirical measurement noise covariance $\overline{R}$ becomes the measurement noise covariance $\mathbf{\Gamma}$. Because it then represents the covariance of the observational noise matrix η of infinite observations $(N_{\text{obs}} \times J)$.

With N_x states, the ensemble matrix X is of size $(N_x \times J)$ resulting in a $(N_x \times N_x)$ covariance matrix $\overline{P}$. The observations z are aggregated in a vector of length N_{obs}. The corresponding measurement noise matrix $\overline{R}$ is square $(N_{\text{obs}} \times N_{\text{obs}})$ and the observation matrix H has dimension $(N_{\text{obs}} \times N_x)$. Appendix A.1.1 illustrates the dimensions of the different matrices and equations.

Fig. 4.2 draws a sketch of how an EnKF evolves for a linear model $f(u, x^j)$. The gray circles indicate the augmented measurements from the filled gray circle that represents the actual measurement. The black circles denote the particles, the filled circles their updated states $x^j_{k|k}$, and the empty circles the forecasted members (index $k|k-1$). To visualize the idea of approximating a probability distribution by particles, the exact probability distributions of both, the members and the measurements, are given as solid lines. The length of the gray arrows represents the covariance matrix $\overline{P}$ – which is a scalar in Fig. 4.2 because it illustrates an ensemble of only a single state. The update pushes each state vector x^j (unfilled black circles) towards its respective measurement z^j, which is indicated by dashed gray lines. The amount of this shifting depends on both, the general error of the ensemble expressed in the KALMAN gain and the individual error of the states. This is why the updated states (filled black circles) are not lying exactly on the dashed lines.

There exists an alternative formulation of the EnKF, which is often used in the field of mathematics. Most works on the convergence, stability, and well-posedness of the filter use it. Appendix A.1.2 presents this formulation. It moves away from the representation above, which stresses the similarities to the classical KF.

Convergence of the ensemble KALMAN filter

Convergence of the EnKF consists of two requirements [BLÖM18]:

1. Collapse of the ensemble to its common mean (necessity), and
2. Convergence of this mean to the ground truth (sufficiency) – which is unknown in most real-world examples.

IGLESIAS ET AL. [IGLE13, Thm 2.1] proved that the ensemble stays within a subspace spanned by the initial ensemble $\boldsymbol{X}_0$ (the so-called "subspace property"). The proof showed that all states of the ensemble are a linear combination of the previous ensemble with a nonlinear weighting – and therefore, of the initial ensemble. They demonstrated that the filter has equal accuracy as conventional least squares methods but does not require any derivatives. In fact, they showed that *"the accuracy is of the same order of magnitude as the best approximation within the linear span of the initial ensemble"* [IGLE13, p. 20]. This assessment also holds for continuous-time formulation [BLÖM18, Lem 3.1]. The rate at which the ensemble collapses depends linearly on the size of the ensemble $\propto J/t$ [SCHI17, Thm 4.2; SCHI18, Cor 3.2; BLÖM18, Thm 4.2] and on the covariance of the observational noise $\boldsymbol{\Gamma}$ [SCHI18, Thm 4.3; BLÖM18, Cor 4.3]. These theorems state that whether the ensemble collapses is only due to the choice of the initial ensemble $\boldsymbol{X}_0$. But the rate at which the ensemble collapses depends on the noise $\boldsymbol{\Gamma}$ and the ensemble size J.

The second requirement, that the common mean converges to the ground truth, does not necessarily follow from the collapse of the ensemble. If the collapse happens too fast, it might be that the ground truth cannot be reached anymore. Especially small ensemble sizes suffer from a slight underrepresentation of variance [EVEN03]. To hold the ensemble from collapsing too fast, the empirical covariance $\overline{\boldsymbol{P}}$ can be inflated artificially with a positive definite matrix (i.e. a symmetric matrix with only positive eigenvalues) [KELL14]:

$$\overline{\boldsymbol{P}}'_k = \overline{\boldsymbol{P}}_k + \boldsymbol{I}\,\alpha_{\overline{\boldsymbol{P}}}^{\;2}, \tag{4.13}$$

with the identity matrix $\boldsymbol{I}$. The empirical covariance is inflated once at each time step before calculating the KALMAN gain, Eq. (4.9). The factor $\alpha_{\overline{\boldsymbol{P}}}$ may be chosen proportional to the standard deviation of the measurement noise covariance matrix [KELL14], e.g. it can be identical:

$$\alpha_{\overline{\boldsymbol{P}}} = \sigma(\boldsymbol{\Gamma}) = \frac{1}{N_{\text{obs}}} \sum_{i=1}^{N_{\text{obs}}} \sqrt{\boldsymbol{\Gamma}_{i,i}}, \tag{4.14}$$

where the (square) measurement noise matrix $\boldsymbol{\Gamma}$ has the dimension of the number of observations N_{obs}. It is reasonable to relate the amount of inflation to the noise level of the measurements. KELLY ET AL. [KELL14] suggested relating the amount of inflation to the error in the initial ensemble, which is more of an academic reasoning than practical advice. Variance inflation is especially of importance for high dimensional spaces and comparably small ensemble sizes because this setting tends to undervalue the covariance matrices [CHAD19]. This forced inflation of the empirical covariance ensures a minimal spread of the ensemble and, thereby, that the search space is properly explored. However, this constant inflation eventually causes the EnKF to fluctuate around the ground truth [KELL14]. Therefore, BLÖMKER ET AL. [BLÖM18] suggested time-adaptive variance inflation to let the ensemble eventually collapse completely. This work inflates the variance by a fraction of the squared standard deviation of the measurement noise covariance, Eq. (4.14). This permanent inflation may sacrifice accuracy but maintains a minimum reactivity if the model coefficients change. The standard deviation $\alpha_{\overline{\boldsymbol{P}}}$ remains constant as long as the measurement noise covariance $\boldsymbol{\Gamma}$ does so.

4.2.2 Considerations on EnKF for identifying a process model in milling

Ensemble size

In its widespread application, EnKF is often used with an ensemble size much smaller than the dimension of the state vector, e.g. in meteorological problems (weather forecasts have million states) [BERG10a; BERG10b]. The work presented here applies the EnKF to a comparably small problem of eight states consisting of the force, the coefficients of the KIENZLE-model for the radial and tangential direction, and of the radial offset of the KLINE & DEVOR-model:

$$\boldsymbol{x} = [\underbrace{F_t \quad F_r}_{\text{force}} \quad \underbrace{k_t \quad k_r \quad m_t \quad m_r}_{\text{force model}} \quad \underbrace{\rho \quad \lambda}_{\text{radial deviation}}\ ]^{\mathsf{T}}. \tag{4.15}$$

The exact process model describes Eq. (4.1).

The number of ensemble members J is also referred to as the ensemble size. A minimum ensemble size is required for the EnKF to work correctly. The larger the ensemble, the better is the approximation of the underlying probability density but also the higher is the robustness towards measurement noise. The computation time scales linearly with the ensemble size $\mathcal{O}(J)$ while the error of the estimation (i.e. the mean of the ensemble) scales with an inverse square root $1/\sqrt{J}$ [EVEN03]. However, the particles in the EnKF collapse again at a linear rate [SCHI17, Thm 4.2; SCHI18, Cor 3.2].

Initial ensemble

According to the subspace property, the EnKF converges within its initial ensemble [IGLE13; BLÖM18]. Therefore, it must be wide enough to contain every reasonable solution. Reviewing the literature, e.g. from the extensive work of KÖNIG ET AL. [KÖNI82], a technologically reasonable range of coefficients for the KIENZLE-model can be derived, Tab. 4.1. The table demonstrates the range of the coefficients from soft material aluminum to the difficult-to-cut nickel super-alloy INCONEL 718. KÖNIG ET AL. [KÖNI82] determined coefficients in turning and for unworn tools.

Table 4.1: *Coefficients of the KIENZLE-model: selected examples and a technologically reasonable range for X5CrNi18-10*

Coefficient Unit	k_t $\mathrm{N/mm^2}$	k_r $\mathrm{N/mm^2}$	m_t –	m_r –
AlZnMgCu 1.5 [KÖNI82, p. 115]	540	56	0.25	0.69
INCONEL 718 [KÖNI82, p. 99]	2112	311	0.29	0.69
X5CrNi18-10 [KÖNI82, p. 95]	1700	350	0.18	0.55
⇒ Upper bound	2400	1000	0.60	0.80
⇒ Lower bound	300	40	0.10	0.25
[ADAM16a, p. 60]	465-673	155-278	0.14-0.28	0.29-0.54

EVENSEN [EVEN03] recommended creating an ensemble by applying perturbations to a best-guess value. In milling, the distribution of the initial ensemble $\boldsymbol{X}_0$ may be arranged around literature values – often obtained from turning. Using these coefficients enhances scientific coherence with regard to new sets of coefficients and may enhance convergence speed. Nonetheless, since the initial ensemble distribution spans the search space, in which the ensemble will converge, a uniform random sampling should be performed instead of a GAUSSIAN sampling. A uniform sampling avoids a bias on the best-guess set of coefficients drawn from the literature. Section 4.3.2 studies the influence of the initial ensemble to the filter.

Imposing constraints

Although the filter converges within the subspace spanned by the initial ensemble – i.e. the mean value of the ensemble converges to a value within this subset. It is not guaranteed that all ensemble members stay within this subset at all times making the filter fragile if the model exhibits singularities or is undefined for certain states. Singularities in the model may eventually affect the KALMAN gain $\boldsymbol{G}_k$ and thereby let the filter deteriorate numerically. Constraints imposed after forecasting, Eq. (4.7), and after the update of the ensemble, Eq. (4.10), keep the filter from running into a model-induced singularity.

CHADA ET AL. [CHAD19] showed that box constraints can be imposed on the EnKF by projecting states, which violate a boundary, back onto the borderline (this method is the "projected NEWTON method"). Imposing constraints in this way may affect the convergence rate of the EnKF because it changes the direction and the step size of the particle in question but it does not affect the overall convergence, i.e. the KARUSH-KUHN-TUCKER (KKT) point of the optimization problem:

$$\min_{\boldsymbol{u}} \frac{1}{2} \left\| \boldsymbol{y} - \boldsymbol{f}(\boldsymbol{u}, \boldsymbol{x}) \right\|^2 . \tag{4.16}$$

If the ensemble is large enough, it is unlikely that many members are affected by box constraints and thus distort the expected behavior of the filter. Eq. (4.16) acknowledges the interpretation of the EnKF solving a (derivative-free least-squares sequential) optimization problem [IGLE13; SCHI17] emphasizing that the optimum does not change if states are projected.

Perturbation of the measurements

Perturbation of the measurements describes how to augment a single measurement $z \to \boldsymbol{Z}$ to create individual observations for every ensemble member. The benefit is not fully understood yet. Presumably, the algorithm benefits from being forced to explore the defined subspace more widely and hence may find a better solution [IGLE13]. The EnKF suffers from degenerative weights [SCHI17]. That is when one of the simulations becomes dominant causing often poor performance of the whole MONTE CARLO simulation, on which the EnKF relies. The filter avoids this pitfall by augmenting the measurements with i.i.d. noise holding all members from falling into a single solution, which is tantamount to degenerating weights in a MONTE CARLO simulation. Augmenting the measurement vector is necessary to maintain diversity within the ensemble. The perturbation must be of zero mean and of the same variance as the measurements. In this way, it reflects the actual measurement (zero mean) and the measurement error (measurement noise covariance $\mathbf{\Gamma}$).

Therefore, this work adds white GAUSSIAN noise to the actual measurement. "White" means of zero mean and equal power over time – or in this case over undeformed chip thickness h. In any case, the EnKF requires the observational (measurement) noise to be known. Assuming knowledge about the measurement noise is reasonable if the sensor can be measured at a known reference load (e.g. no load), and if the noise does not change over time. The deviation from this reference depicts the noise.

Size effect and ploughing at small undeformed chip thicknesses

For an undeformed chip thickness h smaller than a critical thickness, cutting turns into ploughing and mechanistic models loose their validity. Only an ideally sharp cutting edge can avoid ploughing but even a nominal sharp cutting edge exhibits a certain roundness.

The recursive formulation of the EnKF provides certain robustness against the size effect in milling. The size effect manifests itself in drastically increasing model coefficients and pulls the filter towards higher coefficients at every cutter revolution. Thus, the identification benefits from preventing this pull by switching the filter to pull-through when the undeformed chip thickness falls below a critical undeformed chip thickness h_{crit}, which is approximated for non-orthogonal cutting as an empirical threshold h_{TH}.

Although the exponential KIENZLE-model is supposedly less sensitive to the chip size effect, the identification was set to ignore samples where the sum of the undeformed chip thicknesses along the helical cutting edge is less than:

$$\sum_{n=1}^{N_s} h(n) < h_{TH} = 0.05\,\text{mm}. \tag{4.17}$$

This lower limit essentially excludes all samples where ploughing occurs along the entire cutting edge. Individual sections (approximated by disk elements) of the cutting edge can however still be affected. The value was chosen based on the review of the relevant literature: a collection of tables [KÖNI82] and the original works of [KIEN52; VICT56]. Both provide coefficients usually for a range of undeformed chip thickness of $h \in (0.1 - 1\,\text{mm})$. Therefore, measurements at undeformed chip thicknesses much smaller than this range were considered unsuitable for a proper identification introducing the threshold above, Eq. (4.17). It is a reasonable choice, comparing it to other works, Tab. 2.3 in Section 2.2.4, which were all in the range of 0.01-0.6 mm.

4.3 On the identification of a mechanistic force model

4.3.1 Setup for the simulation-based analysis

To study the behavior of the EnKF, this work simulates a cylindrical end milling process of machining stainless steel (X5CrNi18-10). It assumes a solid cemented carbide tool with the parameters listed in Tab. 4.2. In order to mimic realistic conditions and to maintain consistency in this work, the same tool was simulated as was used later in the experiments to validate the MPFC, Chapter 7. Ten revolutions were simulated using the ideal coefficients given in Tab. 4.1. The helix angle β of the tool was approximated by $N_s = 23$ disk elements. A common sampling rate for measurements with a force dynamometer of $f_s = 10\,\text{kHz}$ was assumed. With the cutting velocity given, this resulted in an angular resolution of $\Delta\varphi = 1.3368°$.

Artificial GAUSSIAN noise with different signal-to-noise ratios (S/Ns) was added to the simulated force signals to imitate measurement noise Fig. 4.3. This augmentation enabled a systematic analysis of different methods and – in the case of the EnKF also – different ensemble sizes regarding robustness and convergence.

The measurement noise covariance $\mathbf{\Gamma}$ describes the variance between the modeled, ideal force $\boldsymbol{F}_{act} = [F_{t,act}, F_{r,act}]$ and the artificially created measurements with white GAUSSIAN noise $\boldsymbol{F}_{wgn} = \boldsymbol{F}_{act} + \boldsymbol{\eta}$. The noise $\boldsymbol{\eta}$ (or its

Table 4.2: *Tool and process parameters for the simulation*

Tool geometry			Process parameters		
Diameter	D	10 mm	Total feed	f	0.1 mm
Number of teeth	N_t	2	Cutting velocity	v_c	80 m/min
Helix angle	β	46°	Depth of cut	a_p	2 mm
Rake angle	γ_0	8°	Width of cut	a_e	3 mm

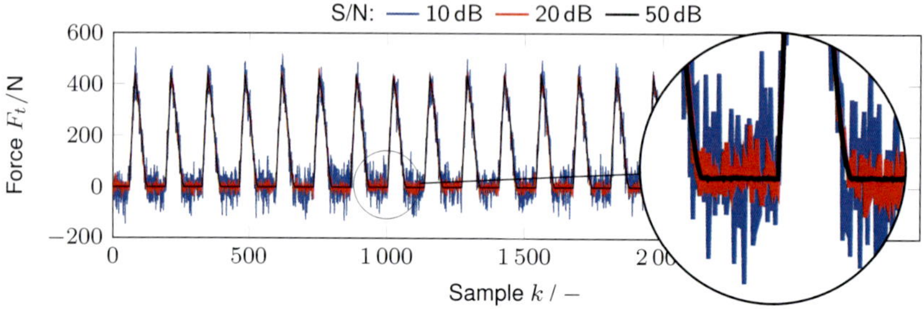

Figure 4.3: *Simulated tangential force signals F_t: ten revolutions with added white,* GAUSSIAN *noise. (Fig. A.1 presents the signals of the radial force component)*

covariance $\mathbf{\Gamma}$) must be quantified once per measurement system. To give an idea of the matrix sizes, their dimensions are written below them.

$$\underset{(1\times2)}{\Delta\boldsymbol{F}} = \underset{(1\times2)}{\boldsymbol{F}_{act}} - \underset{(1\times2)}{\boldsymbol{F}_{wgn}} \tag{4.18}$$

$$\underset{(2\times2)}{\mathbf{\Gamma}} = \underset{(2\times1)}{\left(\Delta\boldsymbol{F} - \overline{\Delta\boldsymbol{F}}\right)^{\mathsf{T}}} \underset{(1\times2)}{\left(\Delta\boldsymbol{F} - \overline{\Delta\boldsymbol{F}}\right)} \tag{4.19}$$

For the sake of conciseness, the illustrations throughout this work are limited to the tangential component of the force and the model (index $(\ )_i = (\ )_t$). The corresponding graphs for the radial components lists Appendix A.3.

To generate a statistically sound database, 500 simulations were performed with different initial ensembles. If a method requires only a single initial state vector, the arithmetic mean was used as the initial guess. In this way, all methods had equal conditions. The initial ensembles were generated as a uniform random set from the range given in Tab. 4.1. This broad range included the KIENZLE-coefficients of all commonly machined metals. The initial ensemble could have been drawn from a smaller range for a specific material, so it presented the worst-case scenario and the most general case.

Determining a proper ensemble size J for the EnKF was difficult because the literature is reluctant to make recommendations. Therefore, a comparative study was performed. Different ensemble sizes used the first J members of a maximum initial ensemble. In this way, the same initial values were used for every ensemble size and combination of measurement noise ensuring consistency.

The results in the following are aggregated in the way that Fig. 4.4 explains. The evolution of each simulation with a distinct initial ensemble (lines on the left side of the subfigures) are aggregated to an area, Fig. 4.4 (a). The transparent area (▬) marks the double standard deviation ($\overline{\Delta F_t} \pm 2\,\sigma(\Delta F_t)$) of all 500 simulations. Assuming a GAUSSIAN distribution, the area represents 95.5 % of the results at a certain sample point k. With regard to time, the presented signals show only samples, where the identification is active (i.e. where no ploughing according to Eq. (4.17) occur), indicated by k'. The solid lines (—) represent the mean value of all simulations.
One simulation represents the evolution of a single EnKF over time k' with a certain size J and initial ensemble $\boldsymbol{X}_0$. Different colors indicate simulations with different ensemble sizes J, Fig. 4.4 (b).

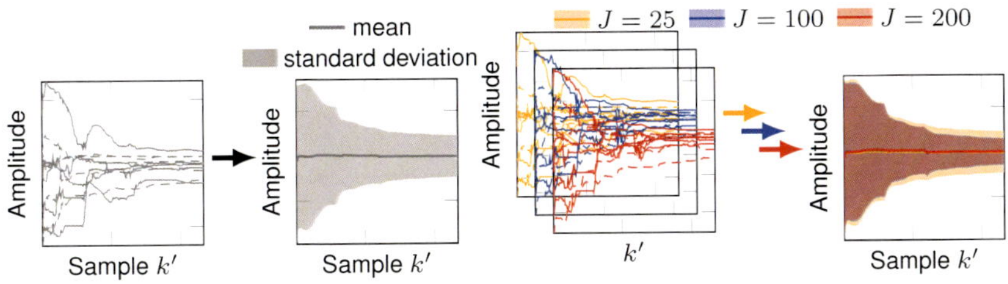

(a) aggregated simulations *(b) aggregated ensemble size*

Figure 4.4: *Sketch of how the simulations are condensed into a single graph*

All calculations were performed with the software MATLAB R2018b from THE MATHWORKS on a computer with an AMD RYZEN7-2700 eight-core CPU (3.06 GHz) and 16 GB DDR3 RAM running MICROSOFT WINDOWS 10.

4.3.2 Comparative study on identification methods for the KIENZLE-model

This section studies the identification of the KIENZLE-model. The tool was assumed to have no radial deviation ($\rho = 0$ in Eq. (4.1)) in order to focus on the force model. Three different identification methods were benchmarked:

- a repeated curve fit (namely BOUNDED OPTIMIZATION BY QUADRATIC APPROXIMATION (BOBYQA)),
- a recursive least squares (RLS), and
- an ensemble KALMAN filter (EnKF).

All belong to the method of instantaneous undeformed chip thickness, Section 2.2.5. The first represents the current state-of-the-art identification method.

It performs discrete, independent identifications building a new model at every revolution of the tool. BOBYQA is a constrained, local, derivative-free optimization algorithm [POWE09] from the open-source software package NONLINEAR OPTIMIZATION (NLopt) [JOHN20]. It presents a compromise between accuracy and execution time in a study on optimization algorithms for the identification of the KIENZLE-model [SCHW19a]. Appendix A.2.2 explains how it works.
RLS is the recursive implementation of a local, unconstrained optimization using explicit derivatives. It is the logical evolution of a curve fit for continuous identification. It had recently been introduced to the field of mechanistic force model identification by [ALTI17; SCHW19b] and proved to be stable only if it was box-constrained [SCHW20].
The EnKF is a novel method for this problem. Based on the idea of a disturbance observer it promises inherent robustness against measurement noise. Its state vector consists of six variables, of which the two forces F_t and F_r can be measured:

$$\boldsymbol{x} = [\underbrace{F_t \quad F_r}_{\text{observations}} \quad \underbrace{k_t \quad k_r \quad m_t \quad m_r}_{\text{force model}}]^{\mathsf{T}}. \tag{4.20}$$

Fig. 4.5 illustrates the evolution of the error in the force F_t for the different identification methods. The curve fit and the RLS are limited to a single set of 500 simulations because they have no size property as the EnKF.
The error is defined as the difference in the (reconstructed) force of the identified coefficients (indicated by a hat) and the ideal coefficients, which are augmented with white, GAUSSIAN noise to mimic force measurements (cf. Section 4.3.1):

$$\Delta F_t = F_{t,sim} - F_{t,ideal} = \hat{k_t}\, b\, h^{1-\hat{m_t}} - k_t\, b\, h^{1-m_t}. \tag{4.21}$$

The methods show an oppositional behavior: while the fit and the RLS are accurate at noise-free signals, the EnKF exhibits poor convergence with a small ensemble size. But in the presence of reasonable noise, the findings switch. The EnKF shows a remarkably fast convergence behavior with low sensitivity to ensemble size. The reason for this speed is that the EnKF benefits from agile ensembles. The agility is directly linked to the measurement noise η since it is considered when calculating the KALMAN gain, Eq. (4.9). The combination of low measurement noise and small ensemble size, in the case here $J = 25$, is especially devastating because the convergence of the mean relates with a square root to the ensemble size [EVEN03]. The EnKF likely has a bad initial estimate (mean of the initial ensemble $\boldsymbol{X}_0$); if it is additionally deprived of its ability to recover through agility, it likely gets stuck around this bad initial solution.

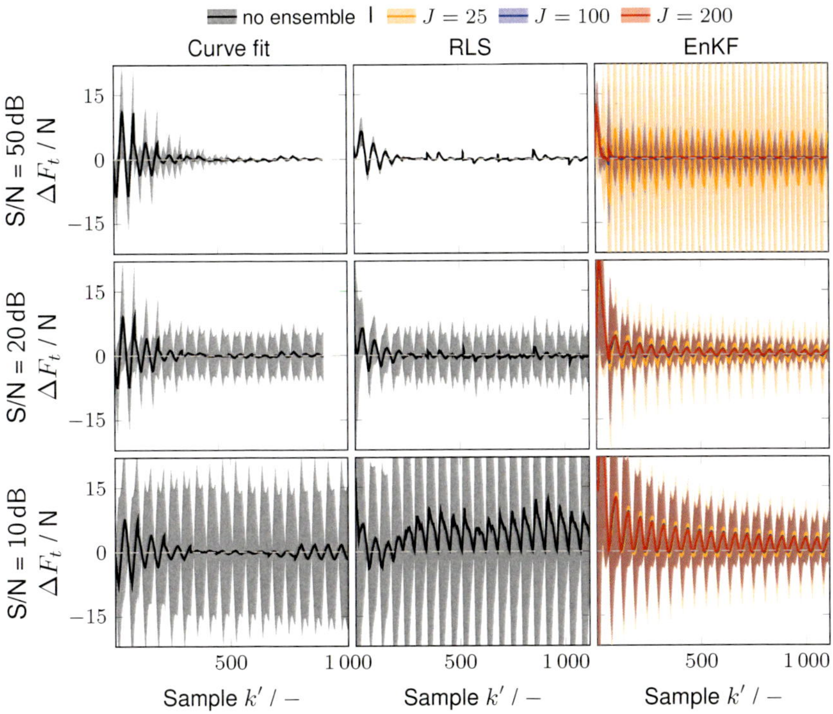

Figure 4.5: *Evolution of the average error (line, —) of the tangential component of the force ΔF_t and its spread ($\pm 2\,\sigma$, shaded areas) for 500 simulations, each with random initial ensembles but identical ideal model coefficients (supplementary figure for the radial component: Fig. A.3, Appendix A.3.1)*

In contrast to this, increasing signal noise (i.e. decreasing S/N; from top to bottom) deteriorated the performance of the curve fit and the RLS. They exhibited a larger spread, but the only difference in the 500 simulations was the initial guess. This finding coincided with studies that found a curve-fit-based identification to be especially prone to disturbing oscillations [WIMM18]. The curve fit was sensitive to noise although it was allowed to run up to 50 iterations in Fig. 4.5, which would jeopardize any requirement regarding computation time. To meet this requirement, the number of function calls must not exceed 20 [ADAM16a; SCHW19a].

Two more important facts can be observed for the EnKF:

- that the ensemble converges (shrinking shaded areas) and

- that the ensemble collapses towards the true value (lines tend to zero error).

These are exactly the two conditions required for convergence [BLÖM18], Section 4.2.1. This behavior should be independent of the measurement noise S/N, although it affects the speed of convergence. At noise-free signals, the artificial perturbation of the measurements in the filter is too small to support convergence (the amount of augmentation is related to the noise $\mathbf{\Gamma}$ of the measurement signals again). This observation suggests that the mobility of the ensemble members is too low to properly explore the search space. The theory addresses the problem of an inert ensemble with "variance inflation," Section 4.2.1 Eq. (4.13), which increases the variance directly. However, too much noise (e.g. S/N $= 10\,\mathrm{dB}$) may hinder convergence speed. Although all three methods perform a derivative-free least-squares sequential optimization, the difference is that the EnKF takes the measurement noise into account making it less prone to its disturbing effect.

4.3.3 Analyzing the convergence of the coefficients

Since the process model is unobservable (i.e. no biunique relation between force and coefficients), there are virtually infinite possible parameter sets for a given force value. Therefore, from the convergence of the error of the force ΔF_t, it does not necessarily follow that the estimated parameters converge to the true values. Simulating the process conditions enables such an analysis.

Fig. 4.6 depicts the evolution of the estimated linear coefficient k_t of all initial ensembles. The findings confirm the ones drawn above in Section 4.3.2 but are more pronounced because the undeformed chip thickness h does not introduce its swelling influence. The larger the ensemble size J, the faster the convergence of the ensemble mean and in the ensemble members, see Section 4.2.1 or [EVEN03; SCHI17, Thm 4.2]. Moreover, the larger the size, the better the approximation of the probability distribution of the state vector x and, consequently, the lower the sensitivity to the initial ensemble (cf. the shaded areas in Fig. 4.6). The analogy of the ensemble size being inertia is misleading. It is not holding the filter from a fast convergence – on the contrary – but it holds the filter from falling for distorted measurement signals. The convergence time is inversely proportional to the ensemble size J with a square root [EVEN03], i.e. the reduction of the error is $40\,\%$ stronger from yellow ($J = 25$) to blue ($J = 100$) than between blue and red ($J = 200$). However, the reduction applies to the overall error of the state vector and is hard to justify on a single coefficient. Small ensemble sizes ($J = 25$, yellow) suffered in particular from a high variation. The high variation became

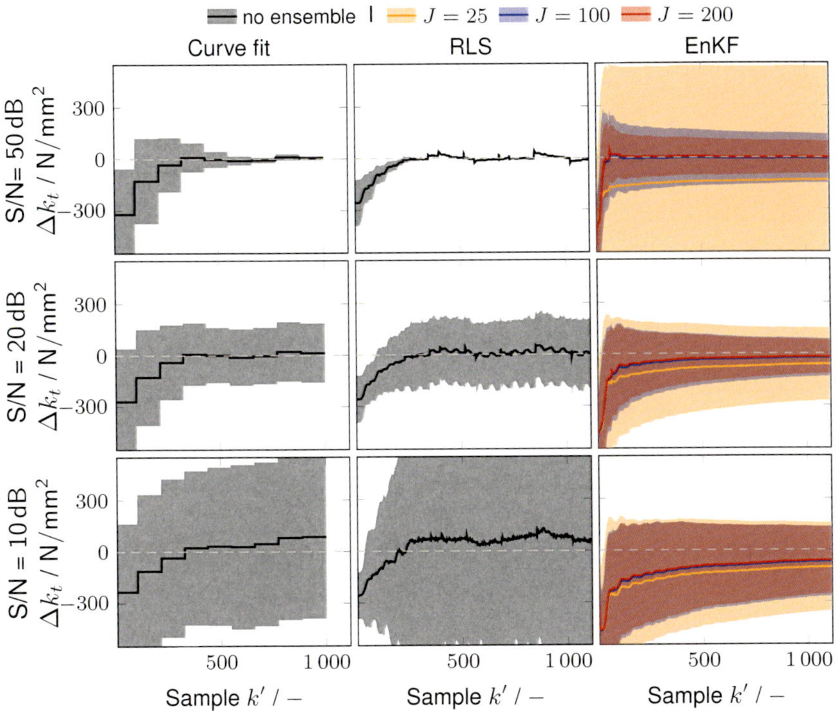

Figure 4.6: *Evolution of the average error (line, —) of the linear coefficient Δk_t and of its spread ($\pm 2\,\sigma$, shaded areas ▬) for 500 simulations each with random initial ensembles but identical ideal model coefficients (supplementary figure for the radial component: Fig. A.4, Appendix A.3.1)*

especially apparent in the evolution of the exponential coefficient m_t, Fig. A.2 in Appendix A.3.1. As before, the RLS demonstrated the highest sensitivity to measurement noise but a very low variance within the initial values. At ideal measurements, only the EnKF with the largest ensemble ($J = 200$) outperformed the RLS with regard to convergence time and stability of the converged value. The curve fit revealed a higher sensitivity to the initial value, visible by larger shaded areas. This dependency was less clear in the force. The discontinuous nature of a curve fit became apparent in the stepwise evolution of the coefficients.

For the sake of conciseness, the rest of this chapter limits the discussion to artificial signals with a noise level of S/N $= 20$ dB, because it best represents signals of a force dynamometer as can be seen later in Section 5.2.2.

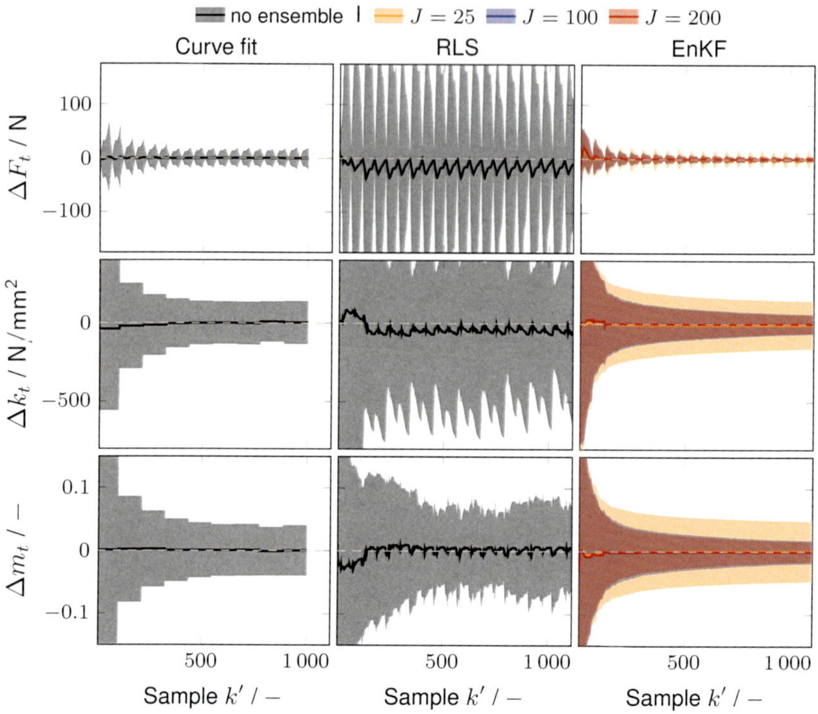

Figure 4.7: *Evolution of the average error (line, —) of the force (top), the linear (middle), and the exponential coefficient (bottom) of the* KIENZLE*-model and their corresponding spreads ($\pm 2\,\sigma$, shaded areas) for 500 random simulations (artificial noise S/N = 20 dB; supplementary figure of the radial components: Fig. A.6, Appendix A.3.1)*

To answer the question of robustness towards force signals in general, i.e. arbitrary model coefficients, the simulation was run again with both, the initial ensembles and the ideal model coefficients, being generated randomly. Fig. 4.7 depicts the results of 500 simulated and artificially distorted force signals.

The EnKF demonstrated a similar robustness and convergence behavior as the nonlinear curve fit, while the RLS showed a large spread with only a minor decrease of it. The graphs confirmed the previous findings for the EnKF that large ensembles favored convergence, which aligns perfectly with the theory [SCHI18; BLÖM18], see Section 4.2.1. However, the improvement seemed to saturate once a sufficiently large ensemble size was reached since the difference between $J = 100$ and $J = 200$ was minimal for the individual coefficients, which were entries of the state vector. The simulations proved that the EnKF reliably converged to the ideal coefficients within a minimum number of samples.

The figures do not show equidistant time steps k but only the time steps k', where no ploughing occurred. For a rough orientation, ten cutter revolutions were simulated, which translate to $t = 10\,T_{rev} \approx 0.24\,\mathrm{s}$ (process parameters given in Tab. 4.2) – provided that the EnKF can be computed within $\Delta t = 1/T_s = 0.1\,\mathrm{ms}$. Section 6.4.3 addresses the issue of efficient implementation and derives the allowed period for the calculation of the EnKF with regard to the overall control system.

4.4 On the identification of a radial deviation model

Considering the radial deviation of the tool is crucial for an accurate model of the instantaneous force. An online identification inherently considers the radial deviation of the entire spindle / tool clamping system. This is important since sometimes these mounting effects dominate the radial deviation of the tool [KRÜG13]. This work uses the model of KLINE & DEVOR [KLIN83], Eq. (2.15), but limits it to the radial offset of the tool, i.e. the parameters ρ and λ, as most of the published works do, Section 2.2.6.

The choice of the subset, from which the initial ensemble is drawn, is important because of the periodicity of the model. While it is theoretically no problem for the EnKF, in practice convergence suffers from this lack of uniqueness. As tools are supposed to exhibit no radial deviation, the initial ensemble should be built around a zero mean $\rho \approx 0$. The uniform distribution for the eccentricity is drawn from a symmetric interval of $\rho \in [-40\,\mu\mathrm{m}, 40\,\mu\mathrm{m}]$. Since this offset is in the same order as the feed per tooth f_t, it represents the extreme case where one tooth cuts while the other cuts twice the intended feed per tooth. Therefore, a reasonable a radial offset is within this range. The direction of the axial offset λ is picked within the limits $\lambda \in [-90°, 90°]$.

The approach is in accordance with the previous one: 500 random parameter sets were drawn from the given boundaries. Each simulation started from a different random initial distribution. In fact, it were the identical model parameters, measurements, and initial ensembles as in Section 4.3.1 – but this time the KIENZLE-coefficients were assumed to be known. Once again, the discrete curve fit and the (box-constrained) RLS should serve as benchmarks. Although the radial deviation model does not distinguish between tangential and radial components, again only the tangential component of the force F_t is illustrated here for consistency.

4.4.1 Identifying the radial offset in polar coordinates

The state vector of the EnKF for the identification of the radial deviation model becomes:

$$\boldsymbol{x} = [\ \underbrace{F_t \quad F_r}_{\text{observations}} \qquad \underbrace{\rho \quad \lambda}_{\text{radial deviation model}} \quad]. \tag{4.22}$$

Circular values have a limited range around which they wrap, making the arithmetic mean not suited for them. One possible approach is to convert polar coordinates to CARTESIAN coordinates, average them there, and convert them back. The $\mathrm{atan2}(x, y)$-function performs the conversion automatically. It calculates a two-argument inverse tangent and is defined as the angle between the x-axis and the vector to the point (x, y) in CARTESIAN coordinates [IEEE-754]. With this function, a circular mean can be defined as:

$$\overline{\alpha} = \mathrm{atan2}\left(\frac{1}{n}\sum_{j=1}^{n} \sin\alpha_j, \quad \frac{1}{n}\sum_{j=1}^{n} \cos\alpha_j\right), \tag{4.23}$$

where α_j is a single element of an n-element vector of angles $\boldsymbol{\alpha}$. The function is sensitive at $\pm 180°$ and may jump the sign.

The curve fit demonstrated an extraordinary ability to minimize the error in the force ΔF_t but maintained a large spread in the model parameters, Fig. 4.8. The EnKF exhibited similar behavior, which might be caused by the representation in polar coordinates and the circular mean. The spread of the RLS exceeded reasonable limits of visualization and showed no sign of convergence. Moreover, in $85.6\,\%$ of the 500 simulations, it deteriorated numerically in such a way that it broke down. Therefore, at the end of the RLS plots, Fig. 4.8 shows only the mean and variance of the remaining 72 simulations. This lack of robustness alone disqualified the method for identification.

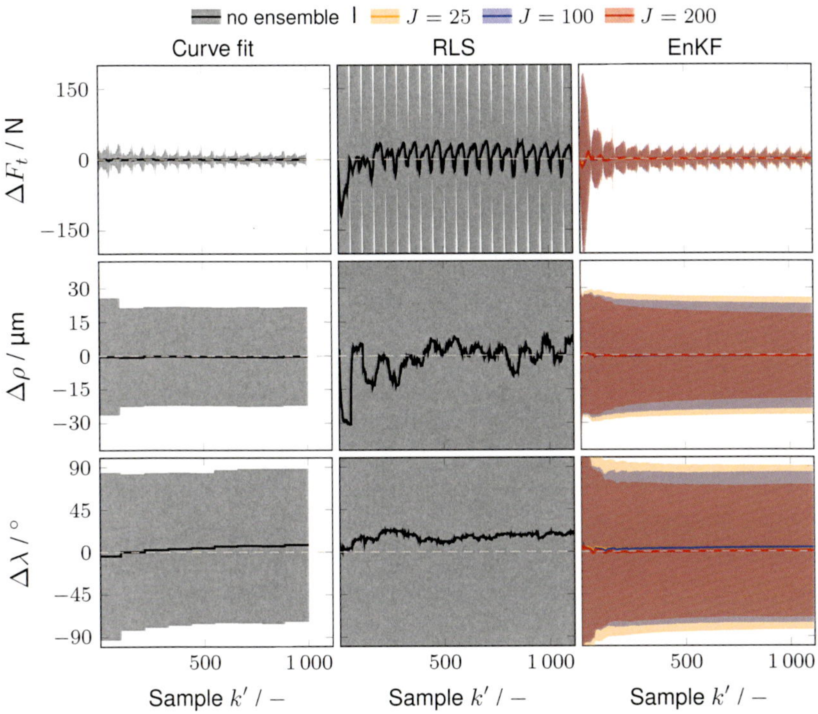

Figure 4.8: *Evolution of the error (average —, $\pm 2\,\sigma$ ▬) in the force (top), the offset (middle), and its direction (bottom) of the* KLINE *&* DEVOR*-model for 500 random simulations (artificial noise S/N* $= 20\,\mathrm{dB}$*; supplementary figure of the radial force: Fig. A.7, Appendix A.3.1)*

4.4.2 Identifying the radial offset in CARTESIAN coordinates

A different approach to address the problem of the discontinuous mean is to work in CARTESIAN coordinates. This requires to conduct a coordinate transformation $(\rho_\mathrm{x}, \rho_\mathrm{y}) \leftrightarrow (\rho, \lambda)$ back to polar coordinates before calling the KLINE & DEVOR-model. The state vector can be represented by two norms ρ_x, ρ_y in orthogonal directions:

$$\boldsymbol{x} = [\, \underbrace{F_t \quad F_r}_{\text{observations}} \qquad \underbrace{\rho_x \quad \rho_y}_{\text{radial deviation model}} \,]. \tag{4.24}$$

Fig. 4.9 presents the results for all identification methods. Now, the EnKF illustrates the expected behavior of gradual convergence in both, the error

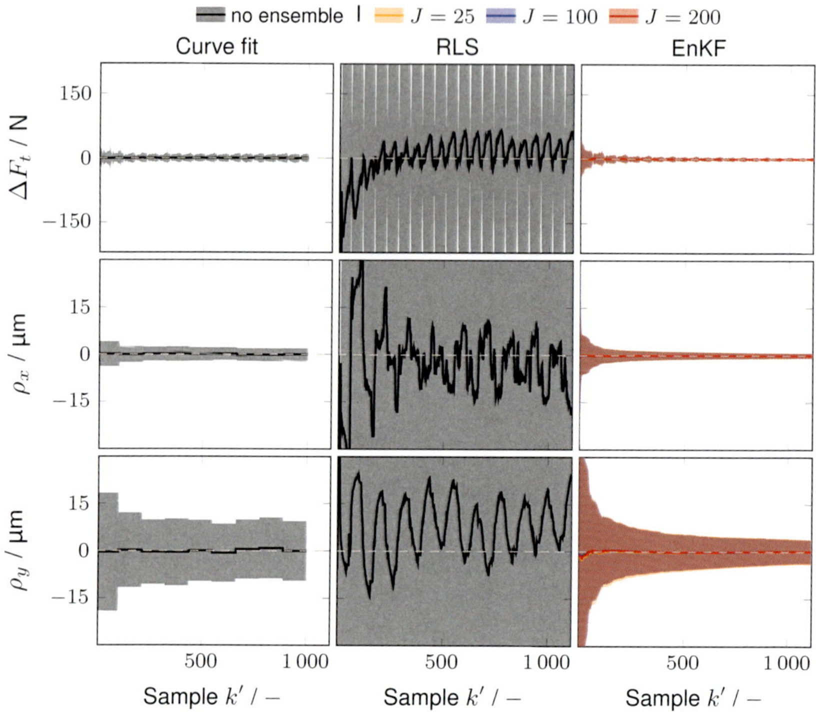

Figure 4.9: *Evolution of the error (average —, $\pm 2\,\sigma$) in the force (top), the offset and direction expressed in* CARTESIAN *coordinates of the* KLINE & DEVOR*-model for 500 random simulations (artificial noise S/N = 20 dB; supplementary figure of the radial force: Fig. A.8, Appendix A.3.1)*

of the force and of the model parameters. Although the model parameters of Fig. 4.8 and Fig. 4.9 are not directly comparable, the error in the force shows a clear advantage of the CARTESIAN formulation. The advantage is acknowledged by the graphs of the curve fit – but not as distinct as the comparison of the EnKFs. Note that it is not possible to infer from a single value of ρ_x or ρ_y to a specific value of the KLINE & DEVOR-model (ρ and λ) because of the coordinate transformation. Therefore, the convergence behavior must be evaluated from both coefficients. The benefit of a CARTESIAN representation of the model parameters for the RLS was limited. The fraction of unstable identifications shrank only by $3.8\,\%$ to $81.8\,\%$ of the simulations (cf. Fig. 4.8). Instability – especially on this scale – was a criterion for exclusion when closing the feedback loop of a controller.

4.5 Analyzing a joint identification of force and radial deviation model

So far, the two models – the force model and the model of the radial deviation – have been analyzed isolated from each other. The respective other parameters were assumed to be known. This approach allowed examining the individual influencing factors. The following combines both identifications: in separate, concurrent identification steps, Eq. (4.26); and as a combined estimation merging the parameters of the models of the force and of the radial deviation into a single filter, Eq. (4.25). The study dropped the RLS as a benchmark because of its poor performance.

$$\boldsymbol{x}_{EnKF} = [\underbrace{F_t \quad F_r}_{\text{observations}} \qquad \underbrace{k_t \quad k_r \quad m_t \quad m_r}_{\text{force model}} \qquad \underbrace{\rho_x \quad \rho_y}_{\text{radial deviation model}} \quad]^{\mathsf{T}}. \tag{4.25}$$

The idea of combining two separate KFs in subsequent steps is known as dual KALMAN filter (DKF) [NELS76; WAN01] – or in the present case dual ensemble KALMAN filter (DEnKF). The overall problem is split into two subproblems, in which a part of the state vector is considered to be known. Usually, this approach exhibits slower convergence but may split a single unobservable problem into two observable subproblems:

$$\begin{aligned}
\boldsymbol{x}_{DEnKF,1,k|k-1} &= \left\{ [\underbrace{F_t \quad F_r}_{\text{observations}} \quad \underbrace{k_t \quad k_r \quad m_t \quad m_r}_{\text{force model}}]_k^{\mathsf{T}} \;\middle|\; \underbrace{[\rho_x,\ \rho_y]_{k-1}}_{\text{radial deviation model}} \right\} \quad \text{and} \\
\boldsymbol{x}_{DEnKF,2,k|k} &= \left\{ [\underbrace{F_t \quad F_r}_{\text{observations}} \quad \underbrace{\rho_x \quad \rho_y}_{\text{radial deviation model}} \quad]_k^{\mathsf{T}} \;\middle|\; \underbrace{[k_t,\ k_r,\ m_t,\ m_r]_k}_{\text{force model}} \right\}.
\end{aligned} \tag{4.26}$$

Since the EnKF does not require observability, feasibility is obsolete as motivation in the case here. Nonetheless, splitting the problem of identifying a process model into the two subproblems studied before (the identification of the KIENZLE-model, Section 4.3; and the identification of the KLINE & DEVOR-model, Section 4.4) may lead to a more benign convergence behavior than a simultaneous identification. Moreover, the DEnKF may be considered as an intermediate step from a fully separate identification towards a joint identification.

The simulations used the same random parameter sets, the same simulated measurements, and the same initial distributions as before. Fig. 4.10 shows the resulting curves of the parameters and the tangential force using the CARTESIAN formulation of the radial offset. The DEnKF was only gradually striving towards zero error. Whereas the single EnKF converged reliably and quickly if the parameters of the model of the radial deviation are expressed as CARTESIAN. The difference was striking: when using the original parameters of the KLINE & DEVOR-model as states, all states showed poor convergence Fig. 4.11 (the coefficients of the KIENZLE-model suggested slow convergence while the polar parameters of the KLINE & DEVOR-model exhibited no convergence). When using CARTESIAN states to represent the radial deviation, all parameters collapsed to their correct solution, causing a small error in the resultant force ΔF_t, Fig. 4.10 (for the figure, the polar formulation of the radial deviation model were used as in Section 4.4.1, so the parameters of the radial deviation model in the state vectors Eq. (4.26) and Eq. (4.25) must be the radial offset ρ and its direction λ).
The curve fit was less affected. It quickly led to a low error in the force and to a gradual reduction of the spread in the parameters. The fit was terminated as before when reaching the maximum number of function evaluations (50). The solution of the previous optimization was taken as the initial guess of the current optimization. In this way, the parameters were loosely linked, eventually converging to the correct solution.

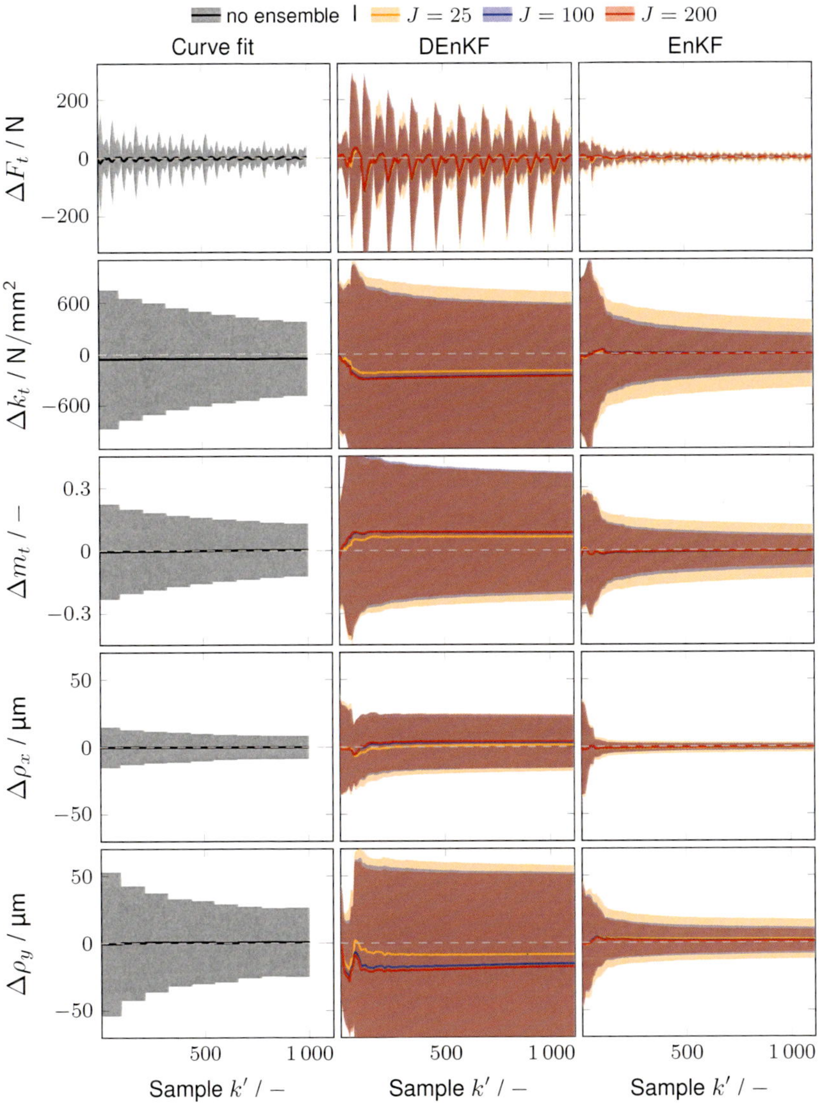

Figure 4.10: *Evolution of the error (average —, $\pm 2\,\sigma$) of 500 random simulations with a simultaneous identification of the* KIENZLE*-model and the* KLINE *&* DEVOR*-model in* CARTESIAN *formulation ρ_x, ρ_y (S/N = 20 dB; supplementary figure of the radial components of the* KIENZLE*-model: Fig. A.10, Appendix A.3.1)*

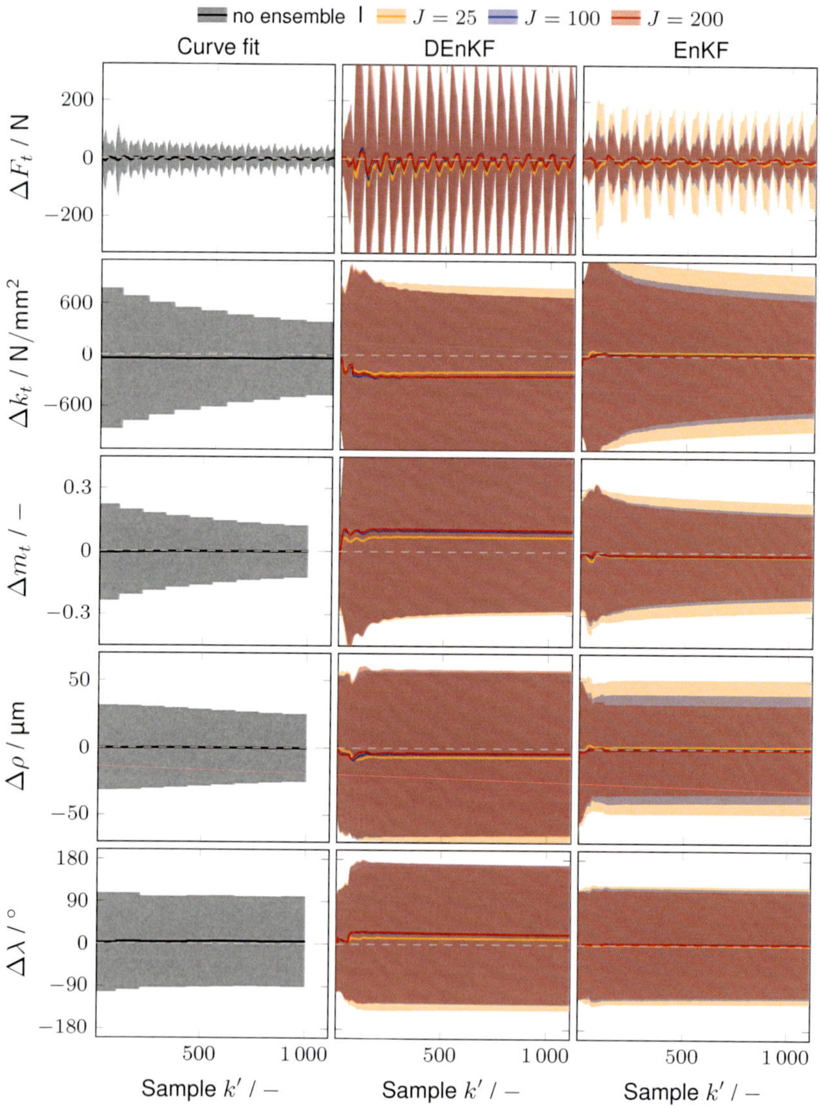

Figure 4.11: *Evolution of the error (average —, $\pm 2\,\sigma$) of 500 random simulations for a simultaneous identification of the* KIENZLE*-model and the* KLINE & DEVOR*-model with the original polar coordinates as to-be-identified parameters (S/N = 20 dB; supplementary figure of the radial force and coefficients: Fig. A.9, Appendix A.3.1)*

4.6 Summary and findings on methods for instantaneous process model identification

A model predictive force controller (MPFC) in milling requires an online identification of a process model. This chapter proposed the ensemble KALMAN filter (EnKF) as a novel, disturbance-observer-based approach for this inverse problem of finding adequate model coefficients to force measurements.

Three methods were compared in simulation: a repeated curve fit representing the state of the art, RLS as a recursive formulation of the previous, and the EnKF. The curve fit built independent models and introduced a time delay to the feedback loop due to its computation and, more significantly, due to a required subsequent smoothing of its output, see [ADAM16a, pp. 105–108]. In contrast to the fit, the estimation of the EnKF and of the RLS provided an immediate and mathematically continuous update of the identification. The mechanistic force model according to KIENZLE [KIEN52] and the radial deviation model of KLINE & DEVOR [KLIN83] were used to model the milling process. On the one hand, was the force model material-specific and thus corresponds to the workpiece-tool combination. On the other hand, the radial deviation coefficients described a characteristic of a specific tool. Ideally, it remained constant over multiple workpieces. Nevertheless, the radial deviation proved to be particularly difficult to identify for the EnKF due to the discontinuous definition of a mean in polar coordinates, in which the radial eccentricity was described. This chapter proposed using CARTESIAN parameters as states in the filter and transforming them to polar coordinates for the model to overcome this problem.

The EnKF exhibited remarkable robustness even in cases where the other methods failed. Often the filter was able to converge in a minimum number of samples – only ten tool revolutions were simulated. The 500 simulated experiments described the worst-case scenario for the identification as the to-be-identified parameters and the initial distribution of the EnKF were selected randomly.

The novel approach of using an EnKF enabled an immediate update of a process model. The recursion of the filter led to a mathematically continuous, smooth identification. The key findings were:

- The EnKF was capable of providing a deterministic, recursive identification.
- The EnKF converged from arbitrary initial distributions in spite of measurement noise – the filter even profited from noise up to a certain extent.

- Convergence time was a matter of computational resources since it was inversely linked to the size of the ensemble J. For process model identification in milling, a choice within the interval $J \in [25, 100]$ was sufficient.

The EnKF stated the first robust online identification of a (mechanistic) process model, and thus, a truly adaptive model of the milling process.

5 Applied model identification with EnKF

So far, the introduced EnKF was examined in simulations of only a few cutter rotations. It was assumed that the measurements were available in the coordinate system of the cutting edge, i.e. that was the tangential force F_t (aka. cutting force) and the radial force F_r (aka. cutting normal force). This chapter leaves these assumptions behind, first transferring the identification method to more relevant coordinate systems (Section 5.1), before applying the technique to real measurement signals (Section 5.2). Progressive tool wear requires to concern oneself with time-variant model coefficients because the coefficients change with increasing tool wear. This systematic procedure ensures that influences and potential sources of error are separated for this new identification technique.

5.1 Identification under coordinate transformation

Mechanistic force models are defined on the cutting edge. However, it is an idealized assumption to measure the force in this coordinate system $(\mathtt{t}, \mathtt{r})$. In fact, it becomes practically impossible if the tool possesses a helix angle $\beta \neq 0$. Research develops piezoelectric coatings to measure force at the cutting edge. Thin-film sensors have been integrated on the tool shaft [MA12; CEN16], on the machine tool table [LUO18a], in the workpiece fixtures [LIU19], or right behind the cutting index [LUO18b; DROS18]. The development is still at an early stage and far from any application.

Forces in a tool-fixed coordinate system can be measured by RCDs. The complex measuring effort, limited transmission rate and bandwidth, as well as high investment costs and restrictions regarding the rotational speed usually prevent the use of such systems. Piezoelectric table dynamometers are more widespread but they are pricey too and limit the machining area to the size of their top plate. They provide force measurements in the coordinate system of the machine tool $(\mathtt{X}, \mathtt{Y})$.

5.1.1 Identification from signals in different coordinate systems

The following examines the behavior of the EnKF in those coordinate systems that are relevant in practice, i.e. in which the process force can be measured:

- force components in the fixed coordinate system F_x, F_y,
- the active force F_a as the resultant of the above, and
- the tangential force F_t, which generates the cutting torque on the spindle.

The last can only be used as additional information to the former measures since it contains no information about the radial component F_r of the process force. Assuming to know the cutting force by measuring the cutting torque $T_t \propto F_t$, the identification can be enhanced. This, in principle, redundant information may solve the problem of information reduction by transformation into a coordinate system, in which forces can actually be measured. Section 5.1.2 discusses this sensor fusion in detail.

The active force F_a is the effective value of the force in the working plane. So it is a reduction of the information compared to the $(\mathtt{t}, \mathtt{r})$-coordinate system of the cutting edge or to the $(\mathtt{X}, \mathtt{Y})$-coordinate system of the machine tool. However, identification based on the active force does not require transforming the modeled force of the cutting edge F_t, F_r into the coordinate system of the machine tool $(\mathtt{X}, \mathtt{Y})$ (or of the table dynamometer respectively). The active force is in particular convenient to measure because it is independent of the angular position of the cutting edge. The active force is responsible for the deflection of the tool, which in turn determines the safety of the process (tool failure) and the resulting workpiece quality (dimensional accuracy). Nonetheless, the literature refrains from identifying model coefficients based on the active force alone. For example, ADAMS [ADAM16a] concatenated the measurements in the machine tool coordinate system and the active force, for his curve-fit-based identification.

Fig. 5.1 summarizes the results. Only combined identifications of the force and radial deviation model in a single EnKF are considered, cf. Eq. (4.25) and Section 4.5. The rather theoretical coordinate system of the cutting edge $(\mathtt{t}, \mathtt{r})$ serves as the ideal reference. As before, the figure condenses the results of 500 random parameter sets of the KIENZLE-model and the KLINE & DEVOR-model– the same sets as before. Each was identified with different initial ensembles of size $J \in \{25, 100, 200\}$. All used the same artificial measurements (S/N $= 20\,\mathrm{dB}$), only transformed to the considered coordinates. The pseudorandom number generator was reinitialized for every identification method to ensure equal randomness.

Three main observations can be drawn right away from Fig. 5.1:

1. The $(\mathtt{X}, \mathtt{Y})$-measurements suggested an almost identical behavior of the identification as in the ideal $(\mathtt{t}, \mathtt{r})$-coordinates.
2. The norm (F_a) could not sufficiently represent the information of the force vector.
3. Considering the tangential component (index $\mathtt{t}$) of the force improved the identification significantly (e.g. $(\mathtt{a}, \mathtt{t})$) – the better, the more uniqueness this information contributed.

A closer look revealed that also the identification based on $(\mathtt{X}, \mathtt{Y})$-measurements benefited from the additional information of the tangential component (third column $(\mathtt{X}, \mathtt{Y}, \mathtt{t})$ of Fig. 5.1). The effect was small but better visible at smaller ensemble sizes. While the mean of the 500 simulations hardly differed between the ensembles and the measurement coordinates, the spread (shaded areas) provided a clearer view. For example, the evolution of the second component of the radial deviation model ρ_y acknowledged the benefit. With regard to the active force (fifth column, $(\mathtt{a}, \mathtt{t})$) the improvement became apparent in the coefficients of the KIENZLE-model, where the error was reduced drastically, which was directly reflected in the force error ΔF_t. Moreover, the results confirmed the general theory, which expected the EnKF to converge – in fact, the filter even did so in the case where active force was the only measurement (column $(\mathtt{a})$). The tendency to converge was clearer from the evolution of the parameters of the radial offset ρ_x, ρ_y, and less in the KIENZLE-coefficients k_t .

Two minor disclaimers must be made with regard to the setup of the analysis. First, forces measured in the machine tool coordinate system $(\mathtt{X}, \mathtt{Y})$ contain less information than if they are obtained on the cutting edge $(\mathtt{t}, \mathtt{r})$, which is an idealized assumption anyway. Shares of the force may cancel each other out at large depths of cut and small helix angles. Therefore, it is utopian to have the same S/N in both cases (the S/N usually is worse, i.e. lower, in fixed coordinates). Nevertheless, for comparison of the coordinate systems, the measurement noise was modeled to be the same in all measurements.
Second, in order to mimic real measurements, it is important to calculate the active force from a model that returns forces in a fixed coordinate system. The information reduction takes place before the active force is calculated when summing up the N_s disk elements, into which the tool is sliced:

$$F_{a,tr}^2 = {F_t}^2 + {F_r}^2 \neq F_X^2 + F_Y^2 = F_{a,XY}^2 \tag{5.1}$$

$$\Leftrightarrow \left(\sum_{n=1}^{N_s} F_{t,n}\right)^2 + \left(\sum_{n=1}^{N_s} F_{r,n}\right)^2 \neq \left(\sum_{n=1}^{N_s} F_{X,n}\right)^2 + \left(\sum_{n=1}^{N_s} F_{Y,n}\right)^2 .$$

Nevertheless, the cutting conditions assumed for the simulations were such that this error did not occur. Excluding the influence of canceling-each-other-out disks demonstrated how drastic the information reduction of the effective value was even without contaminating it with the minor error of the coordinate transformation.

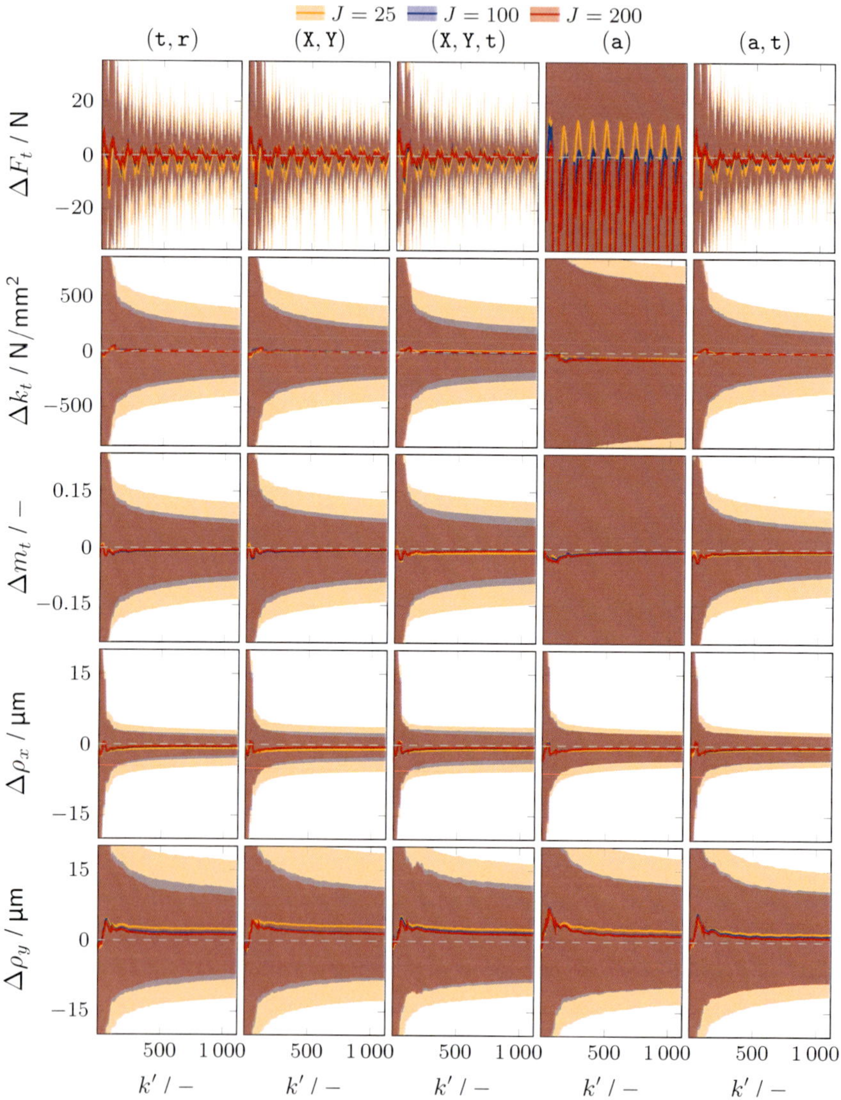

Figure 5.1: *Evolution of the error (average —, $\pm 2\,\sigma$ ▒) for 500 random simulations of an EnKF-based identification of the process model with measurements available in different coordinates systems (S/N = 20 dB)*

5.1.2 Fusing signals of different coordinate systems

Sensor fusion is the art of combining signals from different sources to a piece of joint information. Besides the statistical advantage of relying on multiple sources, the field of research is closely linked to uncertainty reduction, which is usually the intention of sensor fusion.

Nevertheless, HALL & LLINAS [HALL97, pp. 8–9] felt urged to raise a word of warning on this intention: *"In practice, fusion of sensor data may actually produce worse results than could be obtained by tasking the most appropriate sensor in a sensor suite. This is caused by the attempt to combine accurate (i.e., good data) with inaccurate or biased data, especially if the uncertainties or variances of the data are unknown."* This quote emphasizes a proper signal acquisition – an issue that has not lost its significance in modern data analytics, but is nowadays known in a more colloquial form as the *garbage in, garbage out* principle in the machine learning community.

HALL & LLINAS [HALL97] distinguished three levels of data fusion:

1. Raw data level,
2. Feature or state vector level, and the
3. Decision level.

The first can only be used if the same physical quantities are measured. It represents the most simple form e.g. by averaging independent signals of the same quantity as in [POST19]. The second level combines different physical quantities through a model; whereas at the third level, the decision level, every sensor first makes its own decision and a higher-order algorithm merges the independent decisions.

For the case here, the idea is to fuse force signals available in the machine tool coordinate system $(\mathtt{X}, \mathtt{Y})$ with the torque or force signal of the spindle $T_t \propto F_t$. It requires a model to transform the torque (or the cutting force) to the coordinates of the other measurements F_x, F_y. Hence this transformation, it is on the second level of the data fusion classification above.

The most significant difference between the identification of a model and the fusion of sensor signals with a model is the time scope. The identification does not demand a recursive formulation and therefore can be conducted at any rate and on an arbitrary section of the signal. Fusing signals requires calculating the signal fusion at the same rate as the signal acquisition, i.e. in a single time step. In the case here, the charming byproduct is an identified model at every step, which is exactly what the MPFC requires.

To account for the potentially different quality of the force signals, the simulated measurements in $(\mathtt{t}, \mathtt{r})$ direction are augmented with white GAUSSIAN noise of $\mathrm{S/N}_{\mathtt{t,r}} = 20\,\mathrm{dB}$ again. The signals in machine tool coordinates $(\mathtt{X}, \mathtt{Y})$ are assumed to have a ten times poorer quality ($\mathrm{S/N}_{\mathtt{X,Y}} = 10\,\mathrm{dB}$). This reflects the latest research on indirect force measurements based on the motor current of the individual axes (see [MIUR18; MIUR19]) suggests that the fast-rotating spindle provides a significantly better estimate than the slow-moving auxiliary drives. Fig. 5.2 summarizes the results for 500 cases and different ensemble sizes J. The same random parameter sets and the same initial ensembles were used as before.

One can see that the convergence in the force and in the parameters is best when identifying in the coordinate system of the tool $(\mathtt{t}, \mathtt{r})$ with high-quality observations. It shows the identical behavior as above (Fig. 4.10) as they share the same setup. The behavior is much worse in the machine tool coordinate system $(\mathtt{X}, \mathtt{Y})$ with higher signal noise, but it still exhibits convergence in all plots. A significant improvement can be observed if the cutting force F_t is added as a measurement. It becomes especially apparent when comparing the radial deviation parameter ρ_y. The identification with measurements in this mixed coordinates $(\mathtt{X}, \mathtt{Y}, \mathtt{t})$ profits from the redundant information with less noise. The benefit is less pronounced but still visible in the radial components of the KIENZLE-model, Fig. A.11. This observation is remarkable: although the model for the radial force component F_r is independent of its tangential counterpart F_t, its identification (slightly) improves when the tangential force is measured. While the better convergence is obvious in the statistical error of the force ΔF_r – due to a better convergence of the parameters of the radial deviation model, which are shared among both parts of the KIENZLE-model. A slightly more compact spread can be observed in the radial KIENZLE-coefficients k_r and m_r, best notable by a more distinct blue area ($J = 100$) in Fig. 5.2.

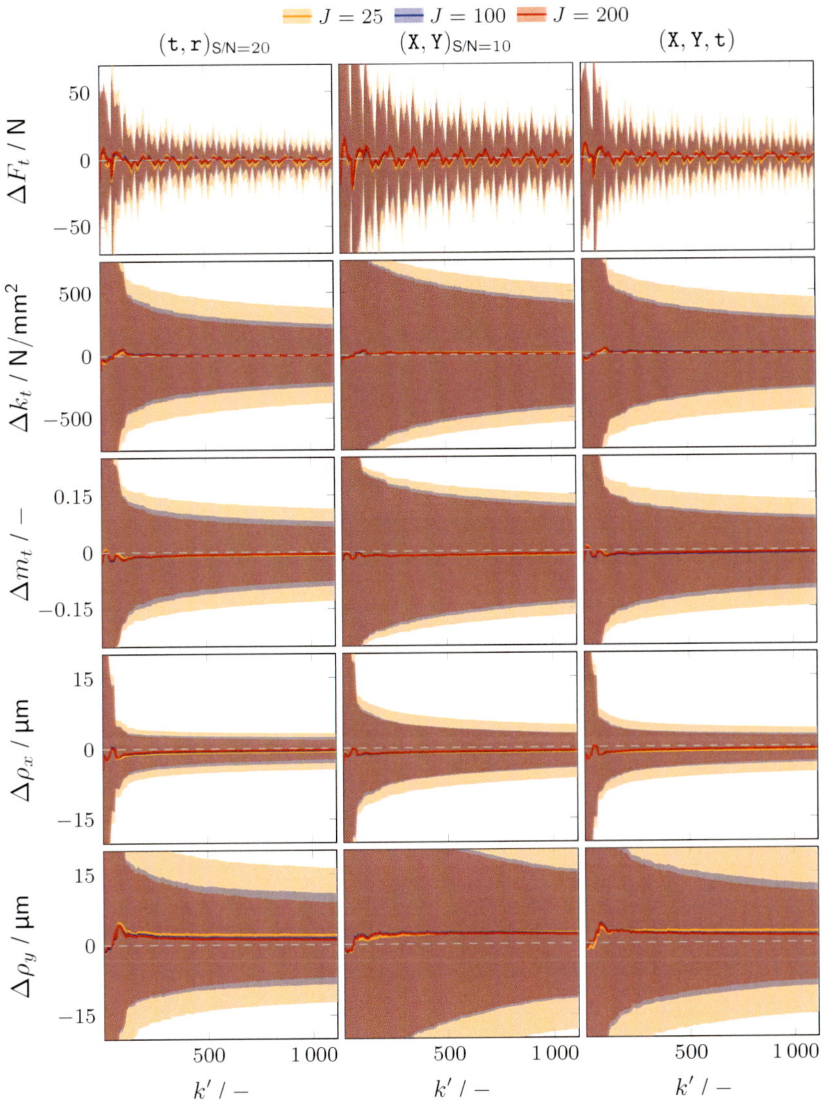

Figure 5.2: *Evolution of the error of 500 random simulations when fusing noisy signals available in the* $(\mathtt{X},\mathtt{Y})$*-coordinate system (S/N*$_{\mathtt{X},\mathtt{Y}}$ $= 10\,\mathrm{dB}$*) with a higher-quality signal available in the tool coordinate system (S/N*$_{\mathtt{t},\mathtt{r}}$ $= 20\,\mathrm{dB}$*); supplementary results of the radial components of the* KIENZLE*-model in Fig. A.11, Appendix A.3.2)*

5.2 Identification on real measurements

This section (Section 5.2) examines the application of an EnKF-based identification on real force measurements. But first, the preconditions are prepared. Section 5.2.1 describes how to measure the position of the cutting edge, while Section 5.2.2 suggests a representative workpiece, which is used for all experiments throughout this thesis. Eventually, the performance of the EnKF is evaluated (Section 5.2.3) and the filter is extended to time-invariant problems (Section 5.2.4).

5.2.1 Measuring the position of each individual tooth

Performing an identification at a single time instant requires the undeformed chip parameters. To infer from the macroscopic engagement conditions to the undeformed chip thickness h and to the width of the undeformed chip b requires the position φ_t of the cutting edge(s) (cf. Eq. (2.11), Eq. (4.1)).

Today's machining centers use encoders to measure the rotary position and speed of their motors. The signals are often encrypted by proprietary protocols, such as SIEMENS DRIVE-CLIQ. Therefore, the original signals of the encoder (sine, cosine, Z-phase analog signals) have to be intercepted and converted to a standard TTL signal, a binary signal for counting. Depending on the exact configuration of this interpolation, the angle of rotation φ can be measured with an accuracy of $0.1°$ and with a resolution of $0.27°$. Usually, an encoder has a second disk with a single increment to indicate a full revolution. This information can be used to reset the relative, up-counted position of the first increment-disk at every revolution generating an automatic modulo operation.

The offset between the angular position of the spindle and the position of the teeth is constant. This difference is defined when clamping the tool in the tool holder, which in turn has a defined position in the spindle to transmit the torque. Thus, the offset is maintained beyond tool changes. It can be determined by measuring it or by reconstructing it. The first can be done by an external measuring machine as it is standard in high-quality and automatic production. A closer-to-process approach would be to measure the position of the individual teeth with a laser-based tool measuring system integrated into the machine tool (intended to calibrate the exact diameter and length of the tool). An automatic procedure would require tapping the signals of the laser in sync with the signal of the spindle encoder. The concept is left for further research.

The second approach reconstructs the position of the teeth from force signals by a curve fit. It is the same procedure as before when determining the

coefficients by an optimization. An observer-based identification, as for the process model, is difficult due to the periodic nature and the drastic nonlinear effect on the force of misplaced positions of the cutting edges.
This work performs a curve fit once before all experiments to determine the angular offset. The distribution of the individual cutting edges on the circumference of the tool is considered to be known for this fit. This "spacing" is often equidistant and usually provided by the manufacturer.

5.2.2 Experimental setup: workpiece, engagement conditions, and measurements

To evaluate the performance of the overall control system, a workpiece is defined that exhibits all manufacturing features, which challenge a force controller in two-dimensional milling: abrupt and continuous changes in the engagement conditions (especially of the width of cut a_e), as well as sections of constant engagement. The abrupt changes challenge the reaction time of the system, while the continuous changes and the sections of no change test the tracking accuracy.

Fig. 5.3 depicts an exemplary tool path used and marks distinct points (1 to 10). It consists of straight cuts with full radial immersion (points 1-2), three-fourth immersion (points 4-5), and half immersion (points 8-9). There is a lateral exit and entry under different angles (points 2 and 4). Abrupt changes in the engagement conditions occur just before / after the entries / exits (points 1, 4, 8 or 2, 5, 9 respectively), where most drastically the depth of cut switches from / to zero. The engagement angles change continuously at every entry and exit until the tool reaches its programmed immersion. The lateral entry / exit at points 2 and 4 respectively prolong this situation. Further, abrupt changes in the engagement angles also occur when the feed direction is changed within the workpiece (points 2, 4, 5, 8).

This corner milling operation serves as a representative for complex workpieces (in terms of the number of workpiece features) with longer tool paths and as a consistent example throughout this work. Using the same example ensures that the results can be compared over chapters. The workpiece material is X5CrNi18-10 (AISI 304); the tool is a two-fluted solid carbide cylindrical end mill (SECO JS512100D2C.0Z2-NXT, Tab. 5.1). Fig. 5.4 shows the corresponding graphs of the engagement conditions at the bottom in a double-axis plot: the depth of cut a_p in black on the left side, and the feed direction α_f and engagement angles φ_{in}, φ_{ex} in gray on the right. Such graphs will be provided for orientation at each figure when discussing signals of this workpiece.

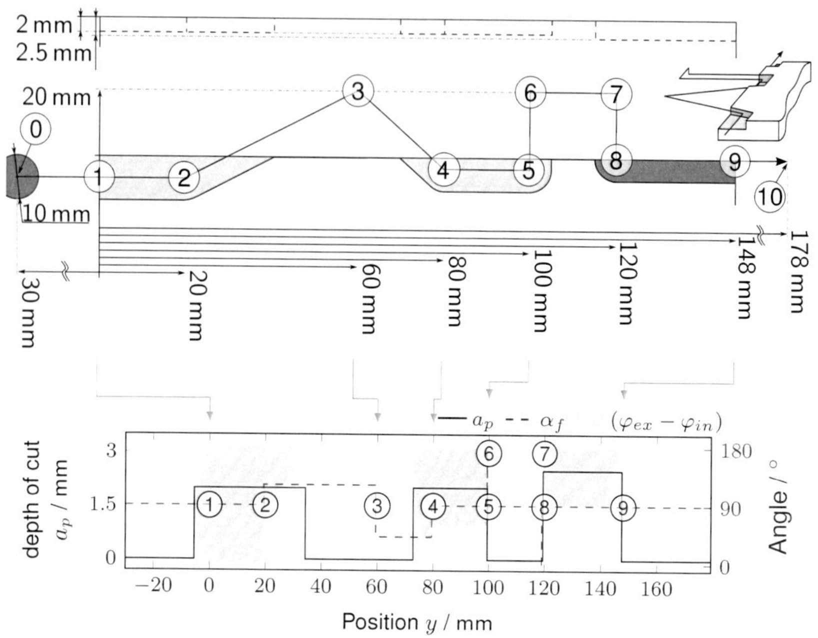

Figure 5.3: *Sketch of the workpiece geometry used for experiments, the tool path, and the corresponding engagement conditions with distinct points marked*

Fig. 5.4 presents the force measurements of manufacturing the example geometry with constant feed ($f_t = 0.0874$ mm or $v_f = 445$ mm/min). Only the regions where the tool is engaged are illustrated for conciseness. The figure presents the measured force as raw signal (black) and as a filtered signal (gray, fourth-order BUTTERWORTH-low-pass filter, $f_{edge} = 300$ Hz) as it is common practice to fight aliasing. It further depicts the simulated force (blue) based on the engagement conditions, the encoder signal of the spindle, and the absolute tool rotation. The angular offset $\varphi_{t,0}$ between the cutting edges and the spindle position and the parameters of the process model (k_i , m_i, ρ, λ) is fitted with the optimization algorithm BOBYQA on the section zoomed in at Fig. 5.5. It is a single cutter revolution from the first region (left column, Fig. 5.4), where the cutter is fully immersed (slot milling). The fit is a least squares fit with a local, constrained optimization starting from an initial guess for X5CrNi18-10 taken from literature [KÖNI82, p. 95]. Details and bounds are shown in Tab. 4.1; for details of the algorithm see Appendix A.2.2. Tab. 5.2 presents the results. The identified coefficients for the KIENZLE-model are close to the ones from literature, Tab. 4.1, [KÖNI82].

Table 5.1: *Tool and process parameters corresponding to Fig. 5.4*

Tool geometry		Process parameter	
Diameter	$D = 10\,\text{mm}$	Feed per tooth	$f_t = 0.1\,\text{mm}$
Number of teeth	$N_t = 2$	Cutting velocity	$v_c = 70\,\text{m/min}$
Helix angle	$\beta = 46°$	Depth of cut	$a_p = 2 - 2.5\,\text{mm}$
Rake angle	$\gamma_0 = 8°$	Width of cut	$a_e = 0 - 10\,\text{mm}$
Material:	solid carbide	Material:	X5CrNi18-10

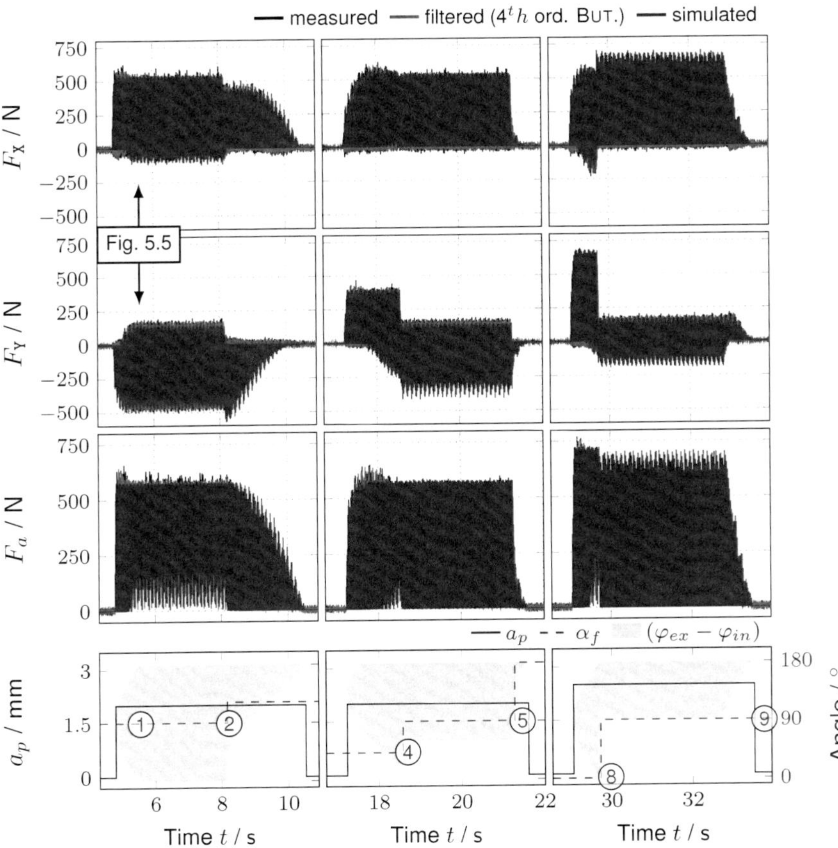

Figure 5.4: *Force signals in machine tool coordinates* $(\mathtt{X}, \mathtt{Y})$ *and corresponding engagement conditions for a process with constant feed* $f_t = 0.0874\,\text{mm}$ *or* $v_f = 445\,\text{mm/min}$*. The circled numbers indicate the points marked in Fig. 5.3.*

Table 5.2: *Fitted parameters of the process model for measurements with a constant feed*

Models	fit on raw sig.		fit on filtered sig.	
Ang. offset $\varphi_0/°$	164.4		142.8	
KIENZLE k_t/N/mm², $m_t/-$	1787	0.185	1797	0.185
k_r/N/mm², $m_r/-$	430	0.549	409	0.563
KLINE & DEVOR ρ/µm, $\lambda/°$	-2.17	-4.3	-2.04	2.2

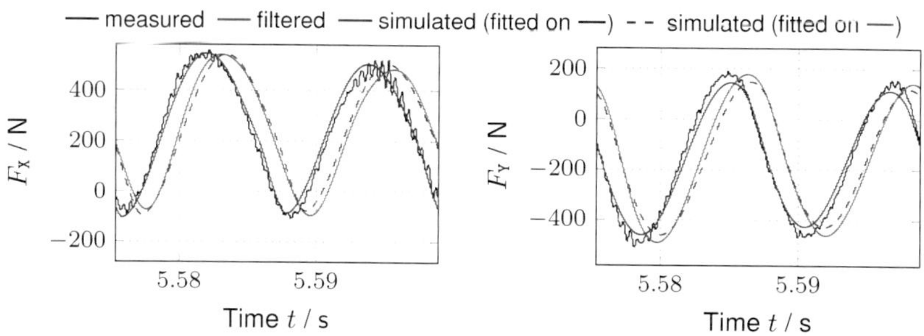

Figure 5.5: *Different forces in machine tool coordinates as an enlargement of Fig. 5.4*

The measurements of this conventionally designed process serve as a benchmark for the control system, Chapter 7. The process parameters (feed f_t and cutting speed v_c) are suggested by the tool manufacturer in its catalog (SECO TOOLS GMBH). The process time is $T_{all} = 36.33$ s, of which the tool was engaged $T_{cut} = 14$ s – the manufacturing without positioning took $T_{prc} = 28.64$ s from the last to the first cut.

Fig. 5.4 also shows the active force F_a (third row), because it will be the control variable later. The most striking deviations occurred when the tool entered the workpiece, where the measurements exhibited an amplification. This gain was caused by the dynamometer/workpiece clamping system as a reaction to the impulse-like excitation. A compensation of this corruption of the measurements was outlined theoretically in REKERS [REKE19] but its online implementation within the cutting process was left for further research.

The force components were measured by a piezoelectric dynamometer (KISTLER 9255C – SN 4940070). It is common practice to set a low-pass filter to fight aliasing. The charge amplifier (KISTLER 5070A10100 – SN 4969962) offered an analog second-order BESSEL filter, while newer products (such as the 5165A) use digital filters and support also a fourth-order BUTTERWORTH filter (details on the filter: [STEP30]). Such filters are widely used in literature, e.g. [NOLZ95a; PARK03; CHAE07; MA12], as they exhibit a sharper roll-off. The filter introduces a time delay. At a sample rate of $f_s = 5\,\text{kHz}$ with a cut-off frequency of $f_{edge} = 300\,\text{Hz}$, a fourth-order BUTTERWORTH filter delays the signal by $1.6\,\text{ms}$. The delay becomes obvious in Fig. 5.5. It has the same effect as if the rotation angle φ would be lagging behind. The fitting (Tab. 5.2) showed an offset of $\Delta\varphi_0 = \varphi_{0,raw} - \varphi_{0,lp} = 21.6°$. With the rotational speed ($n = 2228$ rpm), this lag can be quantified to $\Delta t = \Delta\varphi_0/(n/60\;360°) = 1.6\,\text{ms}$, which corresponds exactly to the theoretical delay. Moreover, the model parameters are almost identical suggesting insensitivity to the low-pass filter. However, the comparison suffers from the higher-order dynamic in the raw signal, especially in the active force F_a. Nonetheless, only the direction of the radial offset λ differs if the signals were filtered or not. Identifying the correct direction is difficult if the radial offset is small because its effect is less pronounced in the force signals (cf. Section 4.4). Consequently, both show high accuracy, Fig. 5.4.

Although used already before when studying the identification in simulation (Chapter 4), the value of the S/N has not been justified yet because it requires real force measurements. The S/N quantifies the amount of noise in a signal. It is defined as the ratio of the average power of the information $\overline{P}_{info}$ that the signal contains over the average power of its noise $\overline{P}_{noise}$:

$$\text{S/N} = \frac{\overline{P}_{info}}{\overline{P}_{noise}} = \frac{\overline{A^2}_{info}}{\overline{A^2}_{noise}} = \frac{\sum_{j=1}^{N} F_{act,j}^2}{\sum_{j=1}^{N} |F_{noise,j} - F_{act,j}|^2}, \tag{5.2}$$

where N represents the number of samples. The ratio of the power can be simplified to their mean amplitudes $\overline{A}_{info}$ and $\overline{A}_{noise}$ respectively. The measure is usually given in decibel: $\text{S/N}_{\text{dB}} = 10\log_{10}(\text{S/N})$. The amplitude of the noisy signal can be split into the signal with the actual information F_{act} and the noise of the signal F_{noise}. Assuming that a curve fit with an unlimited number of iterations obtains the correct values $F_{act} \approx F_{fit}$, the noise level can be quantified to $\text{S/N}(F_{LP}) = 22\,\text{dB}$ and $\text{S/N}(F_{raw}) = 15\,\text{dB}$. The previous chapter demonstrated in Section 4.3.2 that the EnKF-based identification works well at these noise levels.

5.2.3 Analyzing the EnKF on real measurements

This section applies the EnKF as an identification method to real force measurements in the coordinate system of a table dynamometer or of a machine tool respectively $(\mathtt{X}, \mathtt{Y})$. Fig. 5.6 provides the results of the EnKF-based identification on the measurements discussed before in Fig. 5.4. A ground truth exists only for the force, where the force measurements represent the desired reference. The model parameters are presented as absolute values, since none of the approaches can be considered to provide the "correct" solution. Nonetheless, recalling that the EnKF recursively approximates the underlying probability density of the coefficients, the result suggests being trustworthy.

The figure (Fig. 5.6) illustrates different ensemble sizes J and the result of a curve fit for comparison of the model parameters (dashed black line). The fit was performed on ten sections chosen by random from the entire signal to average out effects potentially induced by the local engagement conditions or time. Each section had the length of a single cutter revolution. The results, thus, differ from Tab. 5.2, in which the coefficients were identified in slot milling only (first region). The curve fit had no restriction regarding execution time and, consequently, was not affected by the initial guess.
First, the EnKF-based identification strove towards the results of the curve fit but eventually converged to a different set of parameters. The convergence of the EnKF was the same for all ensemble sizes. Within the shaded areas lied $95\,\%$ of the 500 simulations with different initial ensembles. A clearer ongoing reduction of the spread would have been desired. The remaining spread was due to the fact that the process model did not describe reality completely correctly (cf. the deviations of the slope and the peaks in Fig. 5.5). There were positions of the cutting edges φ_t, where the error became more obvious than at other positions. An identification based on a single time instant could not differentiate between better and less suited positions and has to accept that it is periodically drawn to a worse solution. Pausing the identification for regions of very small undeformed chip thicknesses (Eq. (4.17), Section 4.2.2) improved convergence but represented a compromise between data usage and accuracy. Nevertheless, wrong model parameters had a larger effect on the resulting error of the force at small undeformed chip thicknesses. This led to an oscillation, which made it hard to distinguish accuracy of different ensemble sizes in the force. Moreover, a much longer time was examined here than before (e.g. Chapter 4 or Section 5.1) preventing an analysis on the level of individual tool rotations. The results showed that the EnKF was long-term stable.

As a clearer measure, the error of the active force was illustrated at the peak active force per revolution in the measurements $F_a|_{rev}^{max}$. The ultimate objective was to control this maximum active force $F_{a,max}$ per revolution. Therefore, it was important that the identified model had a small error at the position, where the maximum active force occurred (per revolution).

The ensemble sizes J were smaller than before. Chapter 4 suggested choosing an ensemble size within the interval $J \in [25, 100]$, which was why this

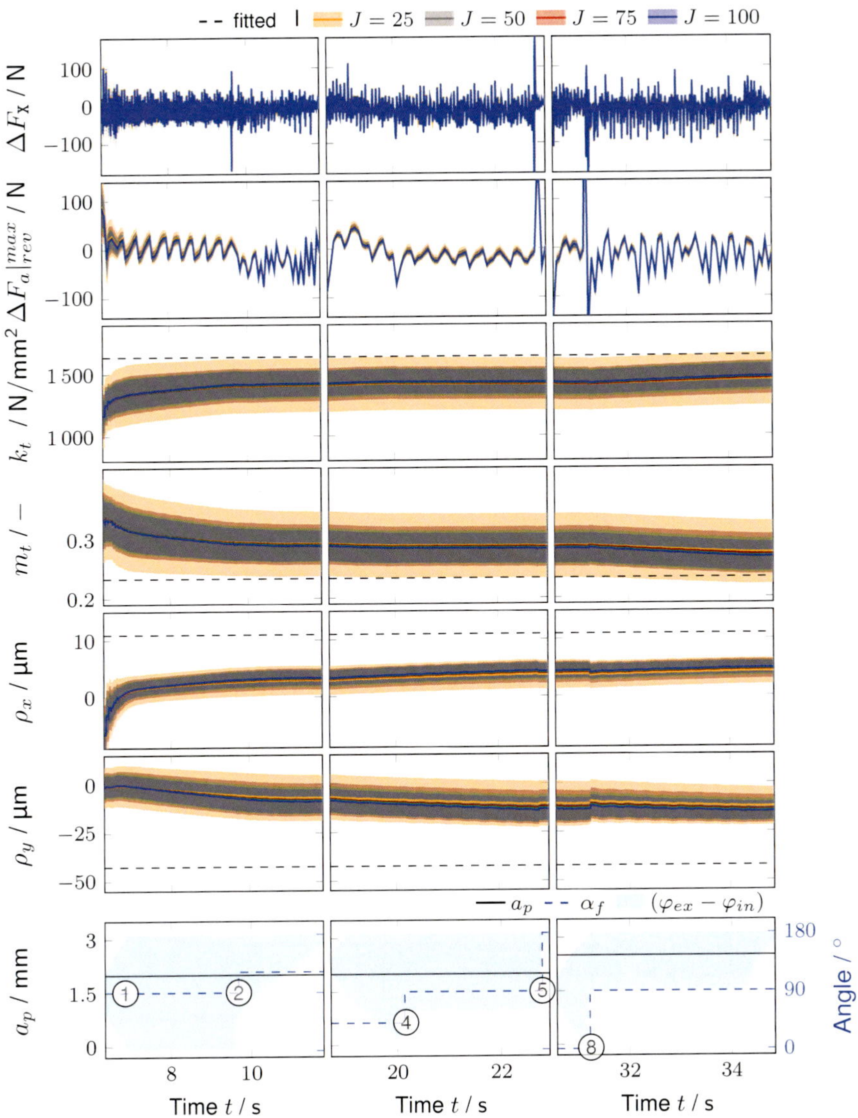

Figure 5.6: *Evolution of the identification on real measurements for 500 random initial ensembles (supplementary figure of the radial force: Fig. A.12, Appendix A.3.2)*

section studied the equidistant set $J \in \{25, 50, 75, 100\}$. Analyzing different sizes evaluated the trade-off between computation $\propto J$ and convergence $\propto 1/\sqrt{J}$ over time k [EVEN03]. Again, 500 simulations were run with the same random initial distributions as before but always on the identical measurements – similar to Section 4.3.2 where the influence of the S/N was examined.

The comparably high error in the force ΔF_{x} in all sections suggested a systematic inaccuracy of the process model rather than of the model parameters. One reason was that no low-pass filter was applied on the simulated force. With a smaller engagement $\Delta\varphi_t < 180°$, the undeformed chip thickness became discontinuous – down milling started with nonzero undeformed chip thickness $h > 0\,\mathrm{mm}$ and exited with a gradually decreasing chip thickness; *vice versa* in the case of up milling. The abrupt entry of the cutting edge in down milling caused an impulse-like excitation, which was more pronounced if the number of engaged cutting edges varied. Step responses were amplified through the transfer behavior of the measurement system and; thereby, they provided biased measurements to the identification, in which the simulated / modeled measurements were not affected by such dynamics. A single-step identification would require a recursive low-pass filter, which was not the scope of this work.

5.2.4 Extending the EnKF to time-variant identifications

The coefficients of mechanistic force models describe a particular wear state of the tool in context of a certain material combination, lubricant, rake angle, and other effects on the process. As wear progresses, the coefficients change gradually – and abruptly if chipping occurs. Ideally, the radial deviation is unaffected if the wear flank does not provoke a drastic change in the macro-geometry of the cutting edge. The problem is that when the ensemble collapses, the EnKF explores a shrunken search space. The filter diminishes its agility making it hard to follow changing coefficients. The EnKF is only partially suitable for time-variant inverse problems. Preliminary works showed that it was improved by inflating the ensemble on a regular basis [SCHW20]. The idea was to inflate the ensemble while maintaining its mean value. In contrast to variance inflation (Section 4.2.1), where the covariance matrix $\overline{\boldsymbol{P}}$ is scaled, ensemble inflation addresses the ensemble members themselves blowing them up like a balloon. This artificial enlargement also inflates the covariance $\overline{\boldsymbol{P}}$ indirectly but at a larger scale than a dedicated variance inflation. Due to this difference in scale, ensemble inflation is performed as a discrete operation that allows the ensemble to converge in the meantime, whereas variance inflation holds the ensemble from a complete collapse. While the classic variance inflation was proposed to increase the reliability of finding the correct solution of a time-invariant problem, this novel approach addresses time-invariant inverse problems. The challenge is to determine how much to inflate the ensemble.

In order to ensure the stability of the filter (and of the to-be-identified model), box constraints were introduced with ensemble inflation (see Section 4.2.2). They kept the model coefficients in a technologically reasonable range (e.g. $m_i < 1$ avoiding that the KIENZLE-model becomes complex). In fact, it was found that a small but frequent inflation was better than a complete reset of the filter from time to time [SCHW20]. As a result, the search space was repeatedly expanded but not completely restored. In conjunction with mean field theory (MFT), only a random subset of the ensemble was inflated in order to obtain a smoother convergence. The results indicated that the EnKF achieved good accuracy if the ensemble was inflated every $k^* = 50$ time steps by a fraction of $\lambda^* = 10$ of the covariance $\boldsymbol{P}_{\boldsymbol{X}0}$ of the initial ensemble: $1/10\ \boldsymbol{P}_{\boldsymbol{X}0}$. In practice, the ensemble inflation was conducted by drawing a new random set of particles with the variance $\overline{\boldsymbol{P}}_k^* = 1/\lambda^*\ \boldsymbol{P}_{\boldsymbol{X}0}$ around the current mean $\overline{\boldsymbol{X}}_k$. The consequence was that the step size k^* must be wide enough so that the initial ensemble $\boldsymbol{X}_0$ converged at least to the size (or variance) of $1/\lambda^*\ \boldsymbol{P}_{\boldsymbol{X}0}$. Otherwise, the ensemble would shrink rather than expand.

The work [SCHW20] presented a general feasibility study with respect to time-varying coefficients of the KIENZLE-model. The study generated arbitrary trends of the coefficients, simulated the force in the $(\mathtt{t}, \mathtt{r})$-coordinate system of the cutting edge assuming constant engagement conditions. Against the background of real measurements, varying engagement conditions, and a significantly longer time horizon, the advice given in the article may had to be adjusted. Therefore, a study on different ensemble inflations with the real measurements from above was conducted. The motivation is broader than just the progression of tool war: the identification shall capture any time-variant effect on the process model.

To compare different combinations of inflation step k^* and inflation fraction λ^*, the deviation area between the actual measurements $F_{\mathtt{X},act}$ and the reconstructed measurements $F_{\mathtt{X},sim}$ was considered as the root mean square (rms):

$$\overline{\Delta F_{\mathtt{X}}}_{,rms} = \sqrt{\frac{1}{K}\sum_{k=1}^{K}\left(F_{\mathtt{X},sim} - F_{\mathtt{X},act}\right)^2}. \tag{5.3}$$

This condensed value should help to grasp different inflation settings and ensemble sizes at a glance. As before, the identification was performed with 500 different initial ensembles. Fig. 5.7 presents the error as box plots. Each box plot shows the 95 %-quantile and the 5 %-quantile as the upper and lower whisker respectively. The box is bounded by the 75 %- and the 25 %-quantile. The dots indicate the maximum and minimum values; the horizontal line in the box the median (i.e. the 50 %-quantile).

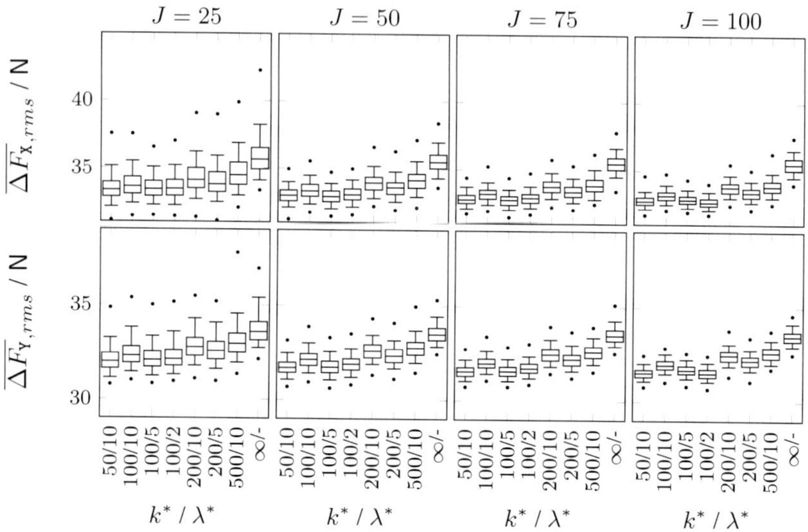

Figure 5.7: *Boxplots of the summed-up rms error between identified force model and measurements for different combinations of inflation steps k^* and inflation fractions λ^*. The parameters of the* KLINE & DEVOR*-model were expressed in* CARTESIAN *coordinates. Fig. A.13 provides the results of the original representation in polar coordinates.*

Fig. 5.7 shows that the identification without any inflation ($k^* = \infty$) was the worst option. The identification seemed to benefit from frequent but small inflation of the ensemble. This finding was supported by previous study on simulated measurements, cf. [SCHW20]. A small but frequent ensemble inflation maintained the agility of the ensemble while allowing it to converge. Severe tool wear was not observed in the experiments ($VB_{max} = 19\,\mu m$). One other reason might be that the smoothing effect of the low-pass filter on the force measurements was only fully effective at sudden changes in force. These changes occurred when the tool was not fully immersed, i.e. from point ② on.

Fig. 5.8 illustrates the evolution of the error and the model parameters for an EnKF, whose particles were inflated $k^* = 50$, $\lambda^* = 10$. Appendix A.3.2 presents the corresponding radial coefficients of the force model in Fig. A.14, and the graphs for the larger inflation with $k^* = 100$, $\lambda^* = 2$ for the sake of completeness in Fig. A.15 and Fig. A.16.

One can see that the model parameters in Fig. 5.8 evolve almost identical to the EnKF without ensemble inflation. It exhibits a marginally larger spread in the coefficients but a more compact error in the force. The slightly smaller error is why the box plot in Fig. 5.7 is more compact. Apart from smaller boxes, the small difference between the graphs with and without ensemble inflation suggests that the model parameters did not change much over this short tool path. The method of repeated ensemble inflation may demonstrate its full benefit only when considering a more severe change in the tool wear state or an inhomogeneous workpiece material.

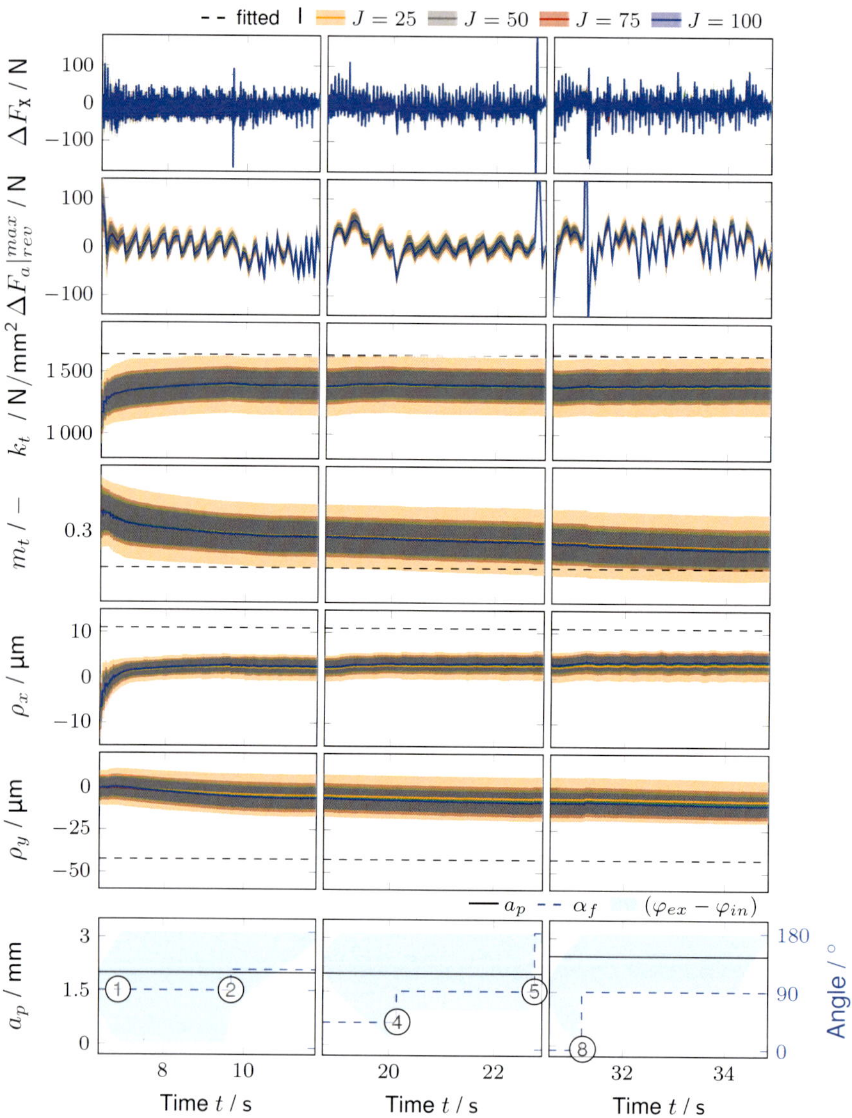

Figure 5.8: *Evolution of the identification with ensemble inflation ($k^* = 50$, $\lambda^* = 10$) on real measurements for 500 random initial ensembles (supplementary figure of the radial force, see Fig. A.14, Appendix A.3.2)*

5.3 Summary and findings on EnKF with real measurements

This chapter applied the theory of an EnKF-based identification to real-world measurements for the first time. It gave up on identifying in the idealized, rotating coordinate system of the cutting edge $(\mathtt{t}, \mathtt{r})$ and, instead, extended the process model to the fixed coordinate system of the machine tool $(\mathtt{X}, \mathtt{Y})$, in which the force measurements were available. Moreover, it prepared the basis for a fusion of (spindle) torque and (CARTESIAN) forces with different S/N. Combining this partly redundant information showed to enhance the overall estimate. Sensor fusion may enable a still accurate identification if the measurements are one day estimated, e.g. by an indirect force measurement rather than measured by a piezoelectric dynamometer.

For instantaneous identification, the positions of the individual cutting edges were required to calculate the undeformed chip parameters. These measures were obtained by measuring the angle of rotation of the spindle and the fixed offset to each tooth. Further, a representative workpiece was defined, which served as a benchmark throughout this work.

Again, the identification proved to be robust to measurement noise. However, high errors occurred at small undeformed chip thicknesses suggesting a systematic error in the force model itself. One reason was that the effect of the low-pass filter of the charge amplifier was not modeled. This provoked large errors between identification and measurements at sudden changes in force. The integration of a recursive low-pass filter must be a priority in future development. Nevertheless, the error at the angular position, where the maximum active force per revolution occurred, was reasonably low ($|F_a|_{rev}^{max}| <$ 40 N). It was this measure what the MPFC ultimately controls.

The process model may change over time due to wear, material inhomogeneity, etc. turning the identification into a time-variant inverse problem. To account for this dependency, the ensemble of the EnKF was inflated from time-to-time in order to maintain a minimum size of the search space. This novel method can be viewed as a special case of the already-known variance inflation, but without holding the ensemble from converging.

This chapter completed the development and validation of an instantaneous identification based on an EnKF. The following chapter analyzes the integration in an online control system with regard to computation, before Chapter 7 evaluates the effect of an EnKF-based identification on MPC for force control in milling.

6 Model predictive force control in milling

Establishing MPC to control the active force in milling requires a larger system in addition to the control algorithm itself. This chapter discusses the system behind it. The chapter explicitly distinguishes between the overall "MPFC system" and the MPC algorithm for clarity. All discussions here are closely linked to computational efficiency, which still is a major concern in the implementation of MPC [SCHW21]. The advent of micro controllers provided the basis for widespread application. However, the more computational power was available, the more demanding the calculations became – either in terms of short sampling times or in terms of model complexity. The work here refrains from using dedicated techniques to reduce the computational effort compared to the original MPC formulation. Examples for such techniques are move-blocking or solving the optimization for all possible states *a priori* ("explicit MPC"), see [SCHW21]. Instead, this chapter trims the main building blocks of the system for efficiency.

The block diagram in Fig. 6.1 outlines the system and may serve as the layout of this chapter. The reference generator (Section 6.1) translates the desired maximum active force $F_{a,des}$ to a reference of the feed velocity. This requires to know the engagement conditions along the tool path and knowing the current parameters of the process model. The MPFC controls the feed velocity v_f using a model of the machine tool. Splitting the system into a part with time delay (the machine tool) and a part with an immediate reaction (the process) emphasizes why also the controller can be split into two parts. Together the reference generator and the MPC-based velocity controller form the force controller, which outputs a velocity command $v_{f,cmd}$. First attempts for a force-based MPC linearized the KIENZLE-model but neglected the influence of radial deviation [STEM19]. Section 6.2 discusses the tuning of the controller: the model of the machine tool, the choice of the horizons, as well as the required resolution of the engagement conditions. When compared to Fig. 2.5 in the state of the art, Fig. 6.1 shows the further development of the MPFC system in recent years.

The engagement conditions are stored as a lookup table together with the position of the tool center point (TCP), upon which they are synchronized online. The TCP must be mapped from the absolute coordinate system of the machine tool $(\mathtt{X}, \mathtt{Y}, \mathtt{Z})$, in which the measurements are available, to the coordinate system of the workpiece, in which the simulation was conducted. This is done by reading the absolute position from the numerical control unit

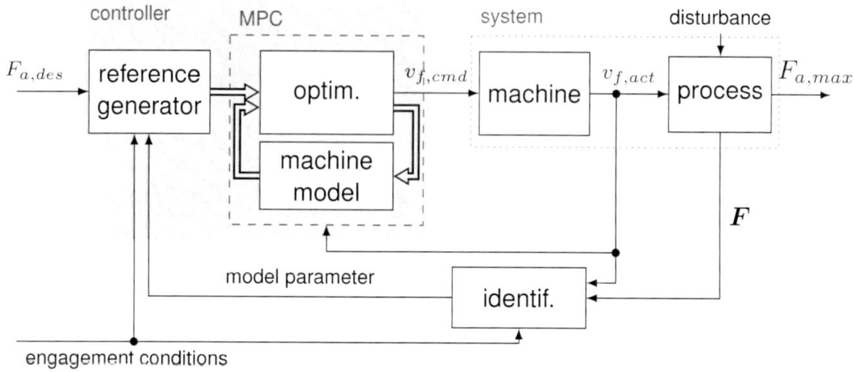

Figure 6.1: *Block diagram of the MPC system for force control in milling*

(NCU) at a defined waiting position, before the process starts, and subtracting it as a constant offset shift.

Although the machine tool is controlled over time, i.e. the MPFC manipulates the feed velocity at a certain sample time, the optimization problem within the MPFC is solved in the one-dimensional space along the tool path, i.e. in position domain. To provide a robust measure of the position along the tool path, the lookup table of the engagement conditions is extended to a column of the traveled tool path. This "measure" is more accurate than integrating the velocity over time because the velocity measure is subject to noise. Further, the in this way obtained position corresponds to the current engagement conditions. The accuracy is only limited by the resolution, with which the engagement conditions are determined (Section 6.2.3).

All experiments conducted in this chapter use the same conditions and process parameters. Therefore, the work refrains from listing all information in every section explicitly. The feed is expressed as a velocity in mm/s, which is what the MPFC system uses. The corresponding – technologically more reasonable measure – feed per tooth f_t is also provided. Both values are exchangeable if the tool geometry and the tool rotation velocity are known. The case given uses a cutting speed of $v_c = 80\,\mathrm{mm/s}$ and the tool Seco JS512100D2C.0Z2-NXT ($N_t = 2, D = 10\,\mathrm{mm}, \beta = 46°$). The work provides both measures to acknowledge the domains production technology and control engineering equally.

6.1 Development of a continuous reference generator

The reference generator is an integral part of the MPFC system and can be considered as part of the controller itself, Fig. 6.1. It closes the feedback loop of the active force F_a in conjunction with the identification of the process model. Since the parameters of the process model change, the feed reference must be adjusted constantly. The essential task is to calculate the inverse of the process model $F_a \rightarrow f_t$. Demanding a reference based on the active force makes it analytically unsolvable. The solution must be obtained by approximating it in an iterative manner. It is important to recall that the reference is a vector and that the complete prediction horizon N_2 of the MPC must be updated. Hence this increased calculation (at every step in time translating multiple time steps to update the reference), execution time is crucial.

The maximum active force can be calculated with the process model (Eq. (4.1), Section 4.1). The model slices the tool into N_s disk elements to approximate the helix angle β:

$$F_a^2\big|_{rev}^{max} = \max_{\substack{j \in \{1,\ldots,N_t\} \\ \varphi_t \in [0°, 360°)}} F_{a,j}^2 =$$

$$\max_{\substack{j \in \{1,\ldots,N_t\} \\ \varphi_t \in [0°, 360°)}} \sum_{i \in (\mathtt{t},\mathtt{r})} \left[k_i \, \frac{b}{N_s} \sum_{n=0}^{N_s-1} \left(f_t \, \sin\varphi_t(n) + \Delta h_t(n)\right)^{1-m_i} \right]^2 . \tag{6.1}$$

This is the maximum of all N_t teeth, because radial deviation produces a unique undeformed chip thickness per slice $\Delta h_t(n)$ and tooth t, Eq. (4.2).

Due to the intermittent nature of the milling process, the (active) force varies within a revolution of the tool. Controlling the force denotes controlling a characteristic force per revolution: the maximum of the active force $F_{a,max}$. It is considered to be critical for tool breakage and for the deflection of the tool (and thus, for the geometric quality of the workpiece). So, the reference generator transforms the reference of the active force $F_{a,des}$ into a reference of the maximum allowed feed per tooth $f_{t,max,des}$.

Instead of calculating the inverse of the force model $F_a(\varphi_t, f_t, \ldots)$ for a full revolution $\varphi_t \in [0°, 360°)$, Eq. (6.1), the idea is to analytically determine the position where the active force becomes maximum $F_{a,max} = F_a(\varphi_t{}^\star, \ldots)$ for the specific engagement conditions of a revolution. This drastically reduces the overall computational effort.

The radial offset constantly points in the same direction and does not rotate with the cutting edge. Therefore, it may be neglected to determine the position, where the maximum active force occurs. Nevertheless, its effect must be considered in the reference generator when translating the force reference to a velocity reference, because the radial offset ρ may even be in the same order as the undeformed chip thickness. For the moment being, the problem of finding the position ${\varphi_t}^\star$ of the maximum undeformed chip thickness collapses to finding the maximum nominal undeformed chip thickness per revolution:

$$\max_{\varphi_t} {F_a}^2 \Rightarrow \max_{\varphi_t} \sum_{n=0}^{N_s-1} \sin\left[\varphi_t - n\,\Delta\varphi_\beta\right]. \tag{6.2}$$

This assumes that the model coefficients k_i , m_i, and $\Delta h_n(\rho, \lambda, \dots)$ do not change significantly within a revolution. This is an approximation since the identification runs faster than the rotating frequency (42.45 Hz).

6.1.1 Determining the maximum undeformed chip thickness per revolution

The aim is to determine the angular position of a cutting edge ${\varphi_t}^\star$, where the undeformed chip thickness h becomes maximum $h({\varphi_t}^\star) = h_{max}$. In general, the first derivative test (necessary condition) states that the slope of a differentiable function becomes zero at extrema:

$$\frac{\partial}{\partial \varphi_t} \sum_{n=0}^{N_s-1} \sin\left[\varphi_t - n\,\Delta\varphi_\beta\right] = \sum_{n=0}^{N_s-1} \cos\left[\varphi_t - n\,\Delta\varphi_\beta\right] \stackrel{!}{=} 0. \tag{6.3}$$

Considering the definition of the engagement angle, Fig. 4.1, and ignoring that the result is periodic, an extremum exists at (in rad)

$${\varphi_t}^\star = \frac{\pi}{2} + \frac{N_s - 1}{2}\Delta\varphi_\beta \qquad \text{for } N_s \in \mathbb{Z}. \tag{6.4}$$

One can intuitively prove that it is an extremum by examining the point symmetry around $\cos\left(\frac{\pi}{2}\right)$, Tab. 6.1.

Although this solution is valid for integers $\mathbb{Z}$ in general, reasonable values for the number of disk elements are positive and greater than zero limiting it to natural numbers $N_s \in \mathbb{N}$.

The second derivative test (sufficient condition) requires the function to have a negative second-order derivative at this point in order to be a maximum. The

Table 6.1: *Illustration of the symmetry around* $\cos(\frac{\pi}{2})$

N_s	${\varphi_t}^\star$	Eq. (6.3)
1	$\frac{\pi}{2} + 0\ \Delta\varphi_\beta$	$\cos\left(\frac{\pi}{2}\right)$
2	$\frac{\pi}{2} + \frac{1}{2}\ \Delta\varphi_\beta$	$\cos\left(\frac{\pi}{2} - \frac{1}{2}\ \Delta\varphi_\beta\right) + \cos\left(\frac{\pi}{2} + \frac{1}{2}\ \Delta\varphi_\beta\right)$
3	$\frac{\pi}{2} + 1\ \Delta\varphi_\beta$	$\cos\left(\frac{\pi}{2} + \Delta\varphi_\beta\right) + \cos\left(\frac{\pi}{2}\right) + \cos\left(\frac{\pi}{2} - \Delta\varphi_\beta\right)$
4	$\frac{\pi}{2} + \frac{3}{2}\ \Delta\varphi_\beta$	$\cos\left(\frac{\pi}{2} + \frac{3}{2}\ \Delta\varphi_\beta\right) + \cos\left(\frac{\pi}{2} + \frac{1}{2}\ \Delta\varphi_\beta\right) +$ $\cos\left(\frac{\pi}{2} - \frac{1}{2}\ \Delta\varphi_\beta\right) + \cos\left(\frac{\pi}{2} - \frac{3}{2}\ \Delta\varphi_\beta\right)$
$\vdots$	$\vdots$	$\vdots$

reasoning is the same as before but exchanges a sinus-function for the cosine in Tab. 6.1, resulting in:

$$\begin{aligned} &\frac{\partial^2}{\partial\varphi_t} \sum_{n=0}^{N_s-1} \sin\left[\varphi_t - n\ \Delta\varphi_\beta\right] = - \sum_{n=0}^{N_s-1} \sin\left[\varphi_t - n\ \Delta\varphi_\beta\right] \\ &\Rightarrow - \sum_{n=0}^{N_s-1} \sin\left[{\varphi_t}^\star - n\ \Delta\varphi_\beta\right] = -1 < 0 \end{aligned}$$

□

The above assumed no restriction on the angular position φ_t. However, the changing engagement conditions along the tool path define where the tool is engaged and cuts material. Considering the current engagement conditions, Eq. (6.4) must be revised to a piece-wise solution. Fig. 6.2 illustrates the cases:

$$ {\varphi_t}^\star = \begin{cases} \frac{\pi}{2} + \frac{N_s-1}{2}\ \Delta\varphi_\beta & \text{for} \quad \varphi_{ex} \geq \frac{\pi}{2} + \frac{N_s-1}{2}\ \Delta\varphi_\beta \text{ and} \\ & \qquad\ \varphi_{in} \leq \frac{\pi}{2} - \frac{N_s-1}{2}\ \Delta\varphi_\beta, \\ \varphi_{in} + (N_s - 1)\ \Delta\varphi_\beta & \text{for} \quad \varphi_{in} \geq \frac{\pi}{2} - \frac{N_s-1}{2}\ \Delta\varphi_\beta, \\ \varphi_{ex} & \text{for} \quad \varphi_{in} \leq \frac{\pi}{2} + \frac{N_s-1}{2}\ \Delta\varphi_\beta. \end{cases} \tag{6.5}$$

6.1.2 Solving the inverse process model

So far, deriving the maximum active force per revolution has been reduced to determining the maximum undeformed chip thickness for certain engagement conditions. Nevertheless, the problem is still the inversion of the process model $F_a \rightarrow f_t$, which needs to be solved numerically, i.e. by iterative approximation.

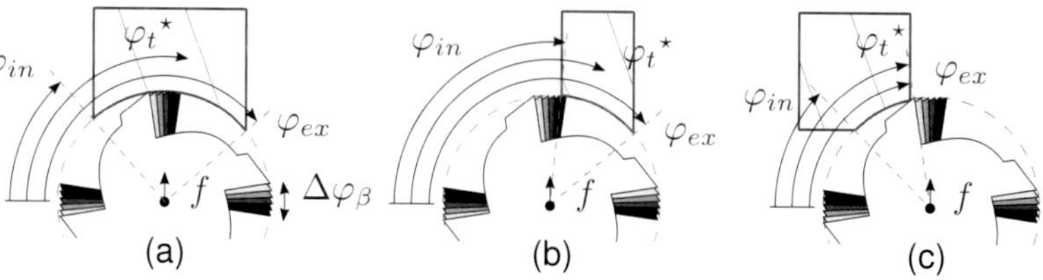

Figure 6.2: *Sketch of the angular position $\varphi_t{}^\star$, where the undeformed chip thickness becomes maximum for different engagement cases*

This section compares different algorithms to solve this problem with regard to execution time:

- SQP,
- BOBYQA, and
- RLS.

All methods are local optimization algorithms. The first two are constrained and derivative-free but RLS is unconstrained and uses the derivative (because it linearizes the problem), Appendix A.2.1. The programming language LabVIEW provides a native nonlinear optimization that uses sequential quadratic programming (SQP). NI, the company developing LabVIEW, provides no details on the theory underlying the implementation. Since the algorithm is native to LabVIEW, in which the MPFC system was built, it is nonetheless included in the benchmark. BOBYQA approximates its nonlinear cost function through a quadratic function, which is adjusted at every iteration. The idea is that the quadratic substitute can be solved much faster than an arbitrary nonlinear problem, Appendix A.2.2. It served as a benchmark before for the identification of the process model and proved to be in particular fast there [SCHW19a]. The BOBYQA-based reference generator was available through a C-based API and it was included as an dynamic-link library (DLL) to LabVIEW.

All three implementations were applied to the engagement conditions of the example workpiece with constant coefficients of the process model assumed. Using a step size or resolution of $\Delta p = 0.075\,\mathrm{mm}$ generated a lookup table with 1551 nonzero entries. The execution time was measured over all 1551 engagement conditions and averaged because the exact timing of algorithms is difficult at submillisecond scale. Tab. 6.2 shows the averaged execution times. Fig. 6.3 illustrates the solution that is the resulting reference of the maximum feedrate $f_{t,max}$.

The stopping criteria of BOBYQA were a maximum execution time of $1\,\mathrm{ms}$ or 25 function calls, whichever was reached first. LabVIEW's SQP algorithm was

Table 6.2: Overview of the execution time of the different implementations for the reference generator

	SQP	RLS 25	BOBYQA (DLL)
total execution time	604 ms	53 ms	84 ms
mean execution time	359 µs	34 µs	55 µs

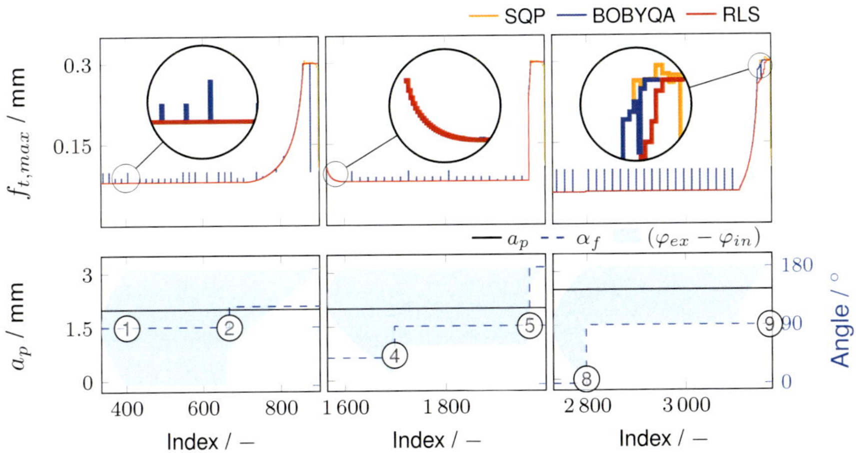

Figure 6.3: *Results of different approaches to solve the inverse of the process model*

stopped already after 20 iterations or when the improvement in the objective function was less than 4 N. Nevertheless, it took the longest for calculation. The RLS was set to perform always a fixed number of 25 iterations. All algorithms started with an initial guess of the feedrate of $f_t = 0.1$ mm.

Although fast (see Tab. 6.2), the zooms in Fig. 6.3 show that BOBYQA exhibited somewhat erratic behavior. This was due to the restriction in execution time, which stopped the optimization after 1 ms maximum. One can see that in the first section (top left axis), the aborted optimizations handed back different solutions although started with identical initial guesses and engagement conditions. In contrast to this behavior, in the third section (top right axis) the solutions were the same for all aborted optimizations. Perhaps such suboptimal solutions could be reduced if the solution at the previous index (or time step) would be taken as the initial guess of the current optimization. Nevertheless, the implementation suggested a lack of reliability for the desired execution time. On the contrary while the SQP reliably obtained the correct solution, its long execution time disqualified the algorithm for integration into the MPFC system.

RLS exhibited a smoother behavior. It relied on the explicit (linearized) derivative of the function. The RLS was implemented in LABVIEW from scratch enabling for full control of the number of iterations that were performed. The custom implementation allowed to perform a fixed number of iterations and thus to reduce the overhead in terms of setting-up the optimization. Originally, the RLS performed a single iteration for a given measurement, but calculating the derivative matrices (JACOBIAN matrix) was the most time-consuming operation per call. Assuming a constant derivative for a given measurement, one can perform multiple optimization steps (or iterations) in a single call. The technique of calling a recursive function on the same measurement is sometimes described as "oversampling". In the present case, the calculation of the derivative was taken out of the loop as a trade-off between accuracy and execution time.

Fast reference generator

A further speed-up was achieved by short-circuit-like operations: the reference generator returned the maximum feed velocity if the depth of cut a_p was zero, i.e. if the tool was not engaged. And it returned the result of the previous position if their engagement conditions matched the current engagement conditions. This focused computation on those points on the tool path, where the engagement conditions change and, with it, the reference of the feed. These tricks reduced the mean execution time per step on the tool path by $60\,\%$ for the example workpiece. To put this in context: updating the entire reference of this example took $22\,\mathrm{ms}$, which almost met a single step of the controller.

Moreover, the calculation of the process model – both in the reference generator as well as in the EnKF – scaled with the number of disk elements N_s, with which the helix angle of the tool is approximated. Following Eq. (4.4), which related it to the instantaneous width of the undeformed chip b, it varied along the tool path. So in order to achieve a constant calculation time the number of disk elements was kept constant based on the expected maximum undeformed chip width b_{max} along the tool path. For all experiments, the number of disk elements was $N_s = 20$.

6.2 Controller design and tuning

The popularity of MPC *"comes in great part from the fact that a suitable model being given, the controller can be easily implemented with a direct physical understanding of the parameters to be tuned and easy constraints handling"* [RICH93, p. 1252]. The effort is shifted from controller design towards modeling [RICH78; GARC89; RICH93]. MPC implicitly determines the control law at every step by solving an optimization problem, which is the main difference to classic controllers [MAYN00].

Nonetheless, MPC offers enormous flexibility regarding its design and tuning [GARR10]. The most significant effect have:

- the model – in particular whether it is a linear or nonlinear model and thus linear or nonlinear MPC,
- the cost function – which is usually kept to the standard formulation,
- the constraints – what and how it is constrained: bound, inequality, or nonlinear constraints, and
- the choice of the solver itself.

The MPC algorithm, Eq. (2.5), exhibits a whole slew of tuning parameters: the horizons (N_1, N_2, N_u), the weight in the cost function (i.e. $\boldsymbol{W_r}$, $\boldsymbol{W}_u$, $\boldsymbol{W_\xi}$), and the time step or sampling time T_s (cf. Section 2.1.5). This is the time, in which the optimization must determine a new value for the manipulated variable u. The weight $\boldsymbol{W}_u$ penalizes the change in the manipulated variable indicating the "aggressiveness" of the MPC algorithm. Nevertheless, the intuitive formulation of the control problem and the fact that MPC is designed in the time domain facilitates its tuning [GARR10]. Regarding the design and tuning of the MPC algorithm, this work follows the work of [STEM20]. The part below discusses the model (Section 6.2.1) and the tuning of the prediction horizons in detail (Section 6.2.2).

The time step, in which the controller or the optimization respectively were run, was set to $T_s = 20\,\mathrm{ms}$ being three times smaller than the time delay of the machine tool and eight times smaller than its settling time as will be discussed in Section 6.2.1. With regard to the cutting velocity v_c, the sampling time allows the MPFC to manipulate the feed twice per revolution. ULSOY & KOREN [ULSO93] almost foresaw this sampling time for the control of cutting processes suggesting $30\,\mathrm{ms}$ in their review article. A sampling time of $T_s = 20\,\mathrm{ms}$ is in accordance with the previous works of the working group [STEM16; ADAM16a; STEM19; AY19a; STEM20].

As solver qpOASES was chosen, the same as STEMMLER and ADAMS and their coworkers used [STEM16; ADAM16a; STEM20]. QUADRATIC PROGRAMMING ONLINE ACTIVE SET STRATEGY (qpOASES) is an open-source optimization algorithm for linear problems, which has *"several theoretical features that make it particularly suited for model predictive control (MPC) applications"* as the project stated [FERR17].

The objective function of the optimization problem is formulated in such a way that it maximizes the position (this entails a maximization of the feed velocity) while ensuring the reference of the feed $v_{f,des}$ as an upper limit [STEM20]. This is the mathematical representation of “be as fast as possible but not faster than”. The optimum often lays at the boundary of the feasible region. It is a desired constraint on the output of the controller and can be relaxed (or softened) to ensure feasibility of the optimization problem. Relaxing a constraint moves it to the cost function where the amount of its violation is penalized. This generates the additional tuning factor $\boldsymbol{W}_{\xi}$, which represents a trade-off between accurate tracking of the reference and smooth control behavior. It is tuned manually until the controller reflects the desired behavior. This work uses $\boldsymbol{W}_{\xi} = 10^4$.
Additionally, hard box constraints on the manipulated variable $v_{f,cmd}$ were considered in the reference generator. They protect the machine and the tool from overload: $f_t \in [0\,\text{mm}, 0.25\,\text{mm}] \Rightarrow v_{f,max} = 21.225\,\text{mm/s}$.

6.2.1 Identifying a machine tool model

The implementation of MPC for force control in milling used a model of the machine tool in the optimization problem, Fig. 6.1. The control performance, i.e. the accuracy of tracking the reference, directly profited from an accurate description of the behavior of the machine tool axes' velocity.
Already in the 1980s the limited response time of the machine tool was acknowledged as a substantial threat to the stability of a classic control system [TU14]. This emphasized that it must be considered in the controller.

The linear axes of machine tools were often modeled as a second-order system, e.g. [FUSS91; ALTI94; ROBE97; ZUPE12b], or less frequent as a first-order system, e.g. [RUBI13]. A second-order system with a time delay T_d can be represented in frequency domain s by a gain K, a damping constant ζ, and the cut-off frequency ω_0 as:

$$G(s) = \frac{K\,{\omega_0}^2}{s^2 + 2\,\zeta\,\omega_0\,s + {\omega_0}^2}\,e^{-s\,T_d}. \tag{6.6}$$

The position controller of a machine tool axis is usually a proportional controller (P gain), while the velocity control loop consists of a proportional gain and an integrating element (PI controller) [WECK06, pp. 152–154]. This structure applies also to an SIEMENS SINAMICS G120 motion controller [SIEM16] of the MAZAK VARIAXIS i600. Previous works have assumed a critically damped system ($\zeta = 1$) with a unity gain $K = 1$ [ADAM16b; ADAM16a], while new measurements suggested that an overdamped system $\zeta > 1$ described the behavior more accurately: $K = 0.9978$, $\zeta = 1.5552$, $\omega_0 = 80.5162$. The MPFC used a discrete state space model. This model can be extended by a multiple of the sampling time to consider the time delay. The step tests drawn in Fig. 6.4 suggested a time delay of $T_d = 60$ ms for the MAZAK VARIAXIS I600. So the machine tool reached 95 % of the commanded value after a settling time of $T_{95\,\%} = 167$ ms.

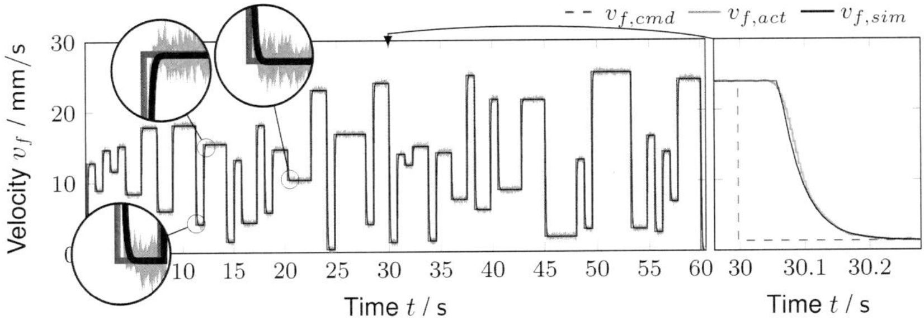

Figure 6.4: *Reaction of the X- and Y-axis of the* MAZAK VARIAXIS I600 *to pseudorandom step changes in the commanded velocity modeled as a second-order system ($K = 0.9978$, $\zeta = 1.5552$, $\omega_0 = 80.5162$)*

Since the feed is manipulated, it requires a model of the velocity behavior along the tool path, which is a mixed-movement of the involved axes. The MAZAK VARIAXIS I600 used is a five-axis, vertical machining center with a SIEMENS SINUMERIK 840D SL controller. The spindle is mounted on the Z-axis, which rests on the X-axis, which again rests on the Y-axis. The linear axes represent a fully functional three-axis milling machine. The double rotary table extends the machine to a five-axis machining center, by adding a degree of rotating (C-axis) and pivoting (A-axis). The identical construction of the linear axes and a similar tuning of their controllers allow to describe them with the same model. This means that the model above can be used to simulate the reaction of the machine tool along a tool path in the X, Y-plane. Fig. 6.4 shows the measurements and a corresponding simulation as a reaction to pseudorandom amplitude modulation of the commanded velocity along a diagonal tool path in the X, Y-plane. The model was identified on measurements of a movement solely in the Y-direction (see Appendix A.3.3 Fig. A.17). The measurements of step changes in the Y-axis and in the X- and Y-axes barely showed a difference. It emphasized the validity of describing both axes with the same model.

Modern motion controllers strive to reduce the attack time by additional feed-forward control. This bypasses the position controller shaping the behavior of the velocity controller [SIEM16]. The purpose is to increase the control dynamic and to reduce the tracking error. The behavior of the machine tool no longer resembles a second-order system but varies with the change of the manipulation variable, i.e. height of the step. Fig. A.18 in the appendix shows the effect of an enabled feedforward control on the same steps as in Fig. 6.4 (in which the feedforward control was disabled). AY ET AL. [AY19b] discussed how to model this nonlinear behavior with support vector regression (SVR) – a method from the field of supervised machine learning. The nonlinear model was linearized successively to be used in linear MPC for force control in milling [AY18; AY19a]. Investigations with regard to online training allowed to adapt the model. They are the subject of further research with regard to the imposed computation. For the work here, the feedforward control was disabled to achieve maximum comparability to the previous works of ADAMS [ADAM16a] and STEMMLER [STEM20].

6.2.2 Choosing control horizons

The prediction horizon N_2, i.e. the number of time steps that the MPC algorithm looks into the future, must be long enough to capture the effect of a change in the manipulated variable u on the system output y. In this way, the maximum length of the manipulated horizon N_u can be estimated by:

$$N_u = N_2 - \left\lfloor \frac{T_d}{T_s} \right\rfloor . \tag{6.7}$$

A larger manipulated horizon would have no effect on the model output. The minimum length is $N_u = 1$. FORBES ET AL. [FORB15] claimed that usability with regard to tuning of MPC is one of the major obstacles for a broad application of MPC in industry. To reduce the complexity of algorithmic tuning, STEMMLER [STEM20] suggested neglecting the difference between the prediction and the manipulated horizon: $N_2 = N_u$. The effect on computation is small if the time delay of the system is small in terms of multiples of the sampling time. Since the machine tool is a dynamic system, where the time delay is only $T_d = 60\,\text{ms} = 3\,T_s$, the recommendation of setting $N_2 = N_u$ is reasonable.

The time delay of the system can be considered in the model or by the lower prediction horizon N_1. The former resembles the more intuitive formulation, which is why the lower prediction horizon was set to $N_1 = 0$. These considerations reduce the problem of finding a suitable set of prediction horizons to the problem of determining only the necessary prediction horizon N_2. Its choice can be estimated with the system model. A second-order system is independent of the step height, i.e. it always reaches 95 % of a value after a characteristic settling time $T_{95\,\%}$, Fig. 6.5. Nevertheless, to deter-

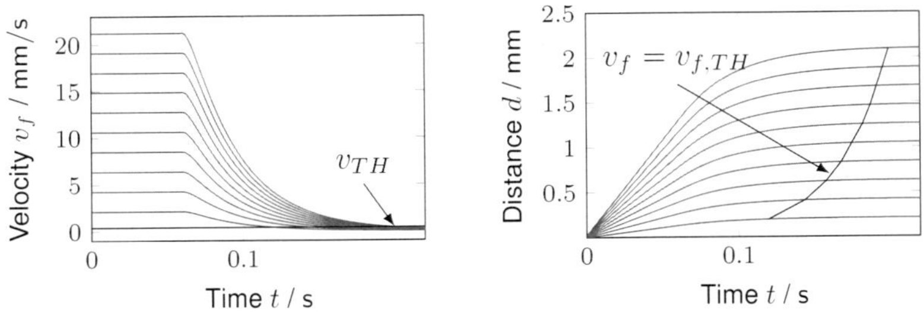

Figure 6.5: *Step change of a second-order system ($K = 0.9978$, $\zeta = 1.5552$, $\omega_0 = 80.5162$)*

mine the maximum traveled distance of the tool, one needs to consider a full stop from maximum speed. Due to its asymptotic convergence, a threshold was defined, below which the machine was considered to have stopped: $f_{t,TH} = 0.005$ mm $\Rightarrow v_{f,TH} = 0.4245$ mm/s. Allowing to move with a maximum feed of $f_{t,max} = 0.25$ mm $\Rightarrow v_{f,max} = 21.225$ mm/s $= 1273.5$ mm/min, this translated to a maximum traveled distance of $d = 2.08$ mm or a traveled time of $\Delta t = 0.1986$ s. With a sampling time of $T_s = 20$ ms, which corresponded to ten time steps, this defined the minimum length of the prediction horizon $N_2 = 10$. The upper bound of the prediction horizon was mainly defined by computational considerations. Stability of the controller profits from longer horizons [SCHW21].

6.2.3 Required resolution of the engagement simulation

The engagement conditions are stored in a lookup table as discrete points along the tool path. The resolution of these equidistantly spaced points causes an unavoidable error in the control of the active force F_a. Allowing a certain error ΔF_a, the required resolution of the engagement conditions can be derived by the inverse of the process model as before (Section 6.1).

A rough estimation can be done by assuming a tool with an orthogonal cutting edge ($\beta = 0°$) and for the maximum undeformed chip thickness $h_{max} = f_t$ by:

$$\Delta F_{a,des} = \sqrt{(b_{max}\, k_t\, {f_t}^{1-m_t})^2 + (b_{max}\, k_r\, {f_t}^{1-m_r})^2}. \tag{6.8}$$

With reasonable model coefficients given, e.g. $k_t = 1700$ N/mm^2, $m_t = 0.18$, $k_r = 350$ N/mm^2, $m_r = 0.55$ for X5CrNi18-10 [KÖNI82, p. 95], a maximum overshoot of $\Delta F_a = 42$ N would require a minimum resolution of $\Delta p_{min} = 2\, f_{t,F_a=42\,\mathrm{N}} = 2.34$ µm. It is twice the feed per tooth because the worst case when searching in the lookup table is the tool being at a position exactly in the middle between two points.

The estimation uses the maximum undeformed chip thickness h_{max} along the tool path and assumes that something between slot milling and half-immersion milling $\varphi_{in} \leq 90°$ is conducted. A resolution of $\Delta p = 0.0023\,\text{mm}$ results in a lookup table with 116,424 rows for the example workpiece (Section 5.2.2). This exceeds any reasonable size – in particular, if compared to the measurements of the position, which are available at 200 Hz resulting in not more than 1500 samples for the given example.

The radial deviation has an extraordinary influence on this analysis as it is typically in the same order as the undeformed chip thickness. If the radial offset exceeds $\rho \geq 1.2\,\mu\text{m}$ (in the direction of the feed), the resolution of the engagement conditions would have to tend towards zero $\lim \Delta p \to 0$. Therefore, it is unreasonable to include the radial deviation in the consideration of the resolution.

The outlined approach resembles a rough estimation. Considering the helix angle β of the tool at the position ${\varphi_t}^{\star}$ of the maximum active force, Eq. (6.4), leads to a more accurate estimation:

$$F^2_{a,\varphi_t{}^\star} = \sum_{i \in (t,r)} \left\{ k_i \, \frac{b}{N_s} \sum_{n=0}^{N_s - 1} \left[f_t \, \sin\left(\frac{\pi}{2} + (\frac{N_s}{2} - n) \Delta\varphi_\beta \right) \right]^{1-m_i} \right\}^2 . \tag{6.9}$$

With the same values as above, the minimum resolution is $\Delta p = 2.37\,\mu\text{m}$ for a maximum overshoot of $F_a = 42\,\text{N}$. This states only a slight improvement in accuracy and a result towards a larger resolution. Therefore, the rough calculation, Eq. (6.8), is sufficient as it represents a conservative estimate on the safe side.

If the engagement conditions change continuously, e.g. if the tool exits under an angle, the error caused by the resolution of the engagement conditions is small. It becomes large at abrupt changes. This can be avoided by including these prominent points. This makes the lookup table unequally spaced but with a maximum resolution of Δp. So, the resolution was set to $\Delta p = 0.075\,\text{mm}$ and the lookup table was extended for prominent points.

To find the corresponding engagement conditions at a certain position, a forward search is performed in the lookup table. From the last position in the table (i.e. the row), the program searches forward to find the closest point to the current measurement. To limit this brute force approach, the maximum step width (in terms of the number of rows) can be roughly estimated by the maximum feed $f_{t,max}$, the revolution speed n, and the sample time $T_s = 10\,\text{ms}$ of the position:

$$N_{ECsync} = (\underbrace{N_t\ f_{t,max}\ n}_{v_{f,max}}\ T_s)/\Delta p = 2.83. \tag{6.10}$$

This estimates the maximum distance that may be traveled provided that the last row index was correct. For robustness, the look ahead N_{ECsync} is set to ten rows and the search starts three rows before the current entry.

6.3 Developing a robust measurement of the feed velocity

The intermittent cut introduces a significant oscillation to the linear axes. This manifests itself in a drastic increase of the noise of the velocity measure when cutting. A closer look at the commanded velocity, which remains constant, shows the severe noise, Fig. 6.6 right. The figure illustrates the signals of the commanded and actual feed of the Y-axis from the NCU at 200 Hz. The problem becomes obvious when considering that standing auxiliary axes exhibit a symmetric velocity profile around zero. This adds a varying offset to the feed velocity when the effective value is calculated. Nevertheless, a potential deviation caused by the process is straightened out at lower-level control loops (velocity and current control loop). Therefore, for practical considerations, the actual velocity can be considered as noisy.

An accurate measure of the feed velocity is paramount because it dominates the calculation of the undeformed chip thickness h. Therefore, noise in the velocity measurement is propagated to the EnKF as an inherent error in the model. The EnKF cannot adequately account for it because the feed is not part of the state vector. Thus, it falsely attributes it to the coefficients deteriorating the identification and the prediction ability of the process model, eventually letting the MPFC fail. Struggling with the same problem, ADAMS [ADAM16a, p. 104] increased the noise covariance of the position estimation within the MPFC. This is not sufficient for an EnKF-based MPFC because both, the identification and the MPC algorithm, rely on an accurate instantaneous measure of the feed velocity.

To obtain accurate velocity information for the upper-level MPFC system, a classic KF is introduced here to fuse the actual velocity with models of the feed axes, which work on the commanded velocity. Appendix A.2.3 provides the theory behind the KF as an optimal linear filter, while the discussion here is limited to the essential setup of the filter: the model and the covariance.

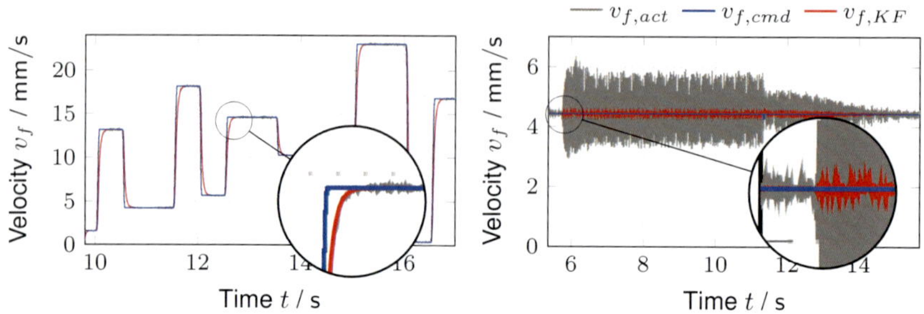

Figure 6.6: *Feed velocity with different measurements techniques (direct and filtered by a KF) in an aircut (left) and in full immersion cutting (right)*

Section 6.2.1 demonstrated that the X- and Y-axes reacted to a manipulated override like a second-order system. Using the same step commands as before but evaluating the behavior of the NCU (the NCU velocity command as well as its measurement) allows to build two further models for the internal behavior of the axes, Tab. 6.3. That the Y-axis exhibits a time delay, but not the X-axis, is plausible since the Y-axis carries the X-axis and, with it, higher inertia.
The difference to the machine tool model previously derived for the MPC algorithm, Section 6.2.1, is that now the reaction to an internal command within the NCU is modeled, while the overlying MPC algorithm requires a model of the reaction to an external command. This explains the significantly longer time delay of the machine tool model for a changed override in Tab. 6.3.

Table 6.3: *Parameters of the second-order system models of the feed axes*

	Gain K	Damping ζ	Nat. freq. ω_0	Delay T_d
X-axis (NCU)	1.0003	0.8253	94.0328	0 ms
Y-axis (NCU)	1.0003	1.6014	90.7206	5 ms
feed (override)	0.9978	1.5552	80.5162	60 ms

A separate KF was built for every axis, with an initial model covariance $\boldsymbol{P} = 10\ \boldsymbol{I}$, a model noise covariance of $\boldsymbol{Q} = 1\ \boldsymbol{I}$, and a measurement noise covariance matrix $\boldsymbol{R} = \boldsymbol{I}\,[100\ 0.1]^\intercal$, in which the velocity information was rated 100 and the position information 0.1. The lower entry on the diagonal of the covariance $\boldsymbol{R}$, the more trusts (i.e. weights) the KF the measured value of the corresponding state. The setting reflects a much higher trust in the position measurement than in the measurement of the velocity. The model is trusted equally for all states. The total feedrate along the path is the effective value of the velocity estimations of all feed axes.

Fig. 6.6 illustrates how different the measured and the estimated feed velocity are at an aircut (left) and at a full-immersion cut (right; this corresponds to the first region of the example workpiece). The measured velocity $v_{f,act}$ of the NCU exhibits a large oscillation with a range of $\Delta v_f = 3\,\text{mm/s} \Rightarrow \Delta f_t = 0.035\,\text{mm}$ when cutting. Further calculations cannot rely on this measure. The blue line indicates the command $v_{f,cmd}$ of the NCU and the red curve denotes the filtered feed velocity $v_{f,KF}$. One can see that the model dominates the filter. Tuning it to further trust the model, it would eventually collapse to a so-called "soft sensor", a sensor whose measurement is processed with a model to generate the output signal.

The filtered feed velocity $v_{f,KF}$ serves as the measurement for the rest of the work $v_{f,act} = v_{f,KF}$. The explicit distinction will not be made outside of this section but it will keep the red color in the figures to come.

6.4 The setup of the model predictive force controller

This section explains the implementation and hardware setup of the current MPFC system. It operates a MAZAK VARIAXIS I600 with a SIEMENS SINUMERIK 840D SL NCU.

6.4.1 Control system communication

That the MPFC system is more than the pure controller becomes apparent when examining the measurements and the communication strings between distinct modules. Fig. 6.7 outlines the communication and their rates. The arrows indicate the direction of the signal flow and the line weight expresses whether it is a scalar value (thin lines) or an array of values (thick lines). Double lines denote signals of multiple time steps. They are sampled at a higher rate than transmitted. This is the case if digitized analog values are buffered in the analog to digital (A/D) converter to meet the read / write limitations of an operating system.
The EnKF calculation can be performed in a loop. The time steps are provided as a batch and the loop processes as many time steps as it can do before the next batch is provided. In this way, the model identification reacts flexibly to the available computational resources. For the controller, the EnKF-based identification presents direct feedback with no delay but it is available at a higher-than-required rate. The identification is conducted at 200 Hz.

Fig. 6.7 conceals the recording of all acquired and processed signals. They were buffered in the RAM and written to a file at 2 Hz. The MPFC system

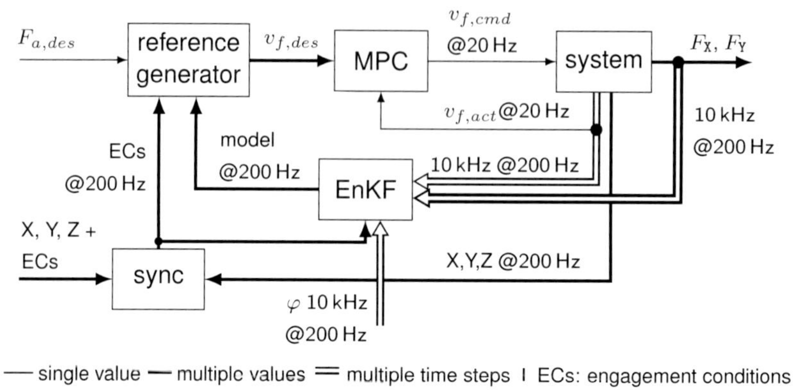

Figure 6.7: *Communication plan of the MPC system illustrating the signals, their sample rate, and at (@) which rate they are available.*

ran on a computer with an AMD RYZEN7-2700 eight-core (3.06 GHz) CPU, 16 GB DDR3 RAM, and the operating system MICROSOFT WINDOWS 10. The position of the tool was read from the internal bus of the NCU. The protocol was PROFINET, from which the SIEMENS AXIS DATA STREAM (ADAS) wrote data at interpolation rate (2 ms) to a CRIO PN BY KUNBUS GMBH module. This was read out at every 5 ms (200 Hz), four times faster than the sample time $T_s = 20$ ms of the controller. The bus-messages were decoded on the field programmable gate array (FPGA)[1] of the compact reconfigurable input/output (CRIO). Using an FPGA minimized the delay introduced by the decoding. It shared a memory block with the real-time part of the CRIO, which transmitted the decoded internal signals to the non-real-time computer, where the MPFC system ran on. The data was transmitted via user datagram protocol (UDP). The protocol transmits data without regard to connection security (additional security layers would lead to delays in the transmission). This was a reasonable risk in a small, local network. Protocols with transmission security through handshakes have a larger overhead and typically wait to accumulate sufficient data to reduce the number of connections.

[1] An FPGA is a logical circuit architecture, which gains its speed from parallel, predefined calculations. It is reconfigurable but not a general-purpose processor as it requires all tasks to be compiled beforehand. In contrast to microcontrollers and conventional processors, the compiled program is a blueprint of a circuit structure for a specific task – not a description of sequential program steps in time. The logical arrays on the chip are reconnected in hardware (they are gate arrays). The functionality is limited but the calculations are deterministic (i.e. hard real-time) and fast through hardware-accelerated parallel computing.

The force was measured by a piezoelectric dynamometer (KISTLER 9255B), which was sampled at 10 kHz. The small charge of the piezoceramic sensors was converted into a voltage signal by a charge amplifier (KISTLER 5070). It also included a low-pass filter of $f_{edge} = 300$ Hz for proper signal conditioning. The rotary positions of the cutting edges with regard to the encoder signal were identified *a priori* through a curve fit.

In summary, the MPFC system required the following measurements:

- force $F_{\mathtt{X}}$, $F_{\mathtt{Y}}$,
- velocity $v_{\mathtt{X},act}$, $v_{\mathtt{Y},act}$, $v_{\mathtt{X},cmd}$, $v_{\mathtt{Y},cmd}$ (for the $v_{f,KF}$),
- position (linear) X, Y, Z,
- position (angular) φ, and
- feed velocity $v_{f,cmd}$ or override (writing access).

Manipulating the overwrite was a common approach to set the feedrate externally, e.g. reported in [TU14]. This was achieved by integrating an analog input module into the electric cabinet of the machine tool and registering it at the NCU so that its value was available as a variable. A "synchronous action" in the NCU was defined, which scaled this value and wrote it to the variable of the feed. Synchronous functions are executed at the same rate as the NCU calculates, i.e. the so-called "interpolation rate". Force, velocity (writing), and the encoder signal were measured through a NI PCIe-6353 measurement board (SN 1673098). This is why they shared the same sampling rate. Even the commanded override was manipulated at this rate although it was updated only at 50 Hz through the MPFC.

The engagement conditions were provided as CSV file with the corresponding position in the workpiece coordinate system. An additional column defined the force reference $F_{a,des}$. Without loss of generality, a constant reference was examined so far [STEM16; ADAM16a; STEM20].

6.4.2 Synchronicity, latency, and delay

It is important that the measurements are in sync, i.e. that they share the same time vector. Related topics are delay and latency. While *latency* describes the time it takes for a signal to travel across two endpoints in a network, *delay* is a more general term, which describes when a piece of information is available with regard to its trigger. This can be the reaction of a system to an input command or – as a special case – buffering e.g. of the A/D converter. Latency and delay between the software modules of the MPFC system are accounted for by the advanced timing operations, which essentially are event-based wait-functions with timeouts.

Synchronization must be ensured between the measurements of the force, the angular position of the spindle, and the position of the tool, which relates to the engagement conditions. With a rotation speed of 2547 rpm, an accumulated time deviation of 1 ms would result in a positional error of $\varphi = 15^\circ$ ruining any point-wise identification. The MPFC system works partly with distributed computing. Signals are preprocessed right on the cRIO. Time is measured based on electronic oscillators (quartz crystals as in clocks) on each device individually. If the frequency deviates between devices, the scale of the time vectors is not equal demanding an adjustment. For a subsequent synchronization, it is sufficient to provide some kind of analog synchronization signal, e.g. random steps, to all devices. For online processing, the clocks must be synchronized once beforehand. Both approaches allow a synchronization of distinct points. There exist dedicated network protocols, which may synchronize the time stamps several times per minute ensuring synchronicity of the signals among multiple devices. In combination with a low-latency network, a synchronized time step makes signals available immediately across multiple devices mimicking a single measurement device. One example protocol is the time-sensitive network (TSN) developed by the IEEE (working group 802.1). TSN is a promising technique for the MPFC system in the future. The previous system was less sensitive to minor delays as it performed a cross-correlation before the curve-fit-based identification [ADAM16a, p. 53; STEM20, p. 160]. This work only checked if a delay occurred between the internal signals and the external signals (force and encoder) by comparing a pseudorandom step signal. Generated on the FPGA, it was transmitted together with the internal signals via UDP over LAN and, at the same time, via an analog line to the main computer. The analog transmission is free of delay and, therefore, in sync with the external signals. This double routing enabled to quantify a potential delay of the internal signals. Experiments where the signals were not in sync was repeated.

6.4.3 Real-time implementation

Although the EnKF-based identification is a single-step estimation, the number of particles (that are the ensemble members) poses a computational burden on the identification. Besides limiting the ensemble size J, how the filter is integrated into the MPFC system is an important factor for the computation time. For this reason, this section studies different ways of efficiently integrating the EnKF in the MPFC system. The EnKF was developed and studied in MATLAB, but the MPFC system ran in LabVIEW. The findings are somewhat broader on an abstract level providing general advice on integrating MATLAB to LabVIEW.

General processing pipeline

Although MATLAB can be directly integrated into LabVIEW through a script node, it is not the most efficient way. The node calls the MATLAB runtime environment in the background, which is limited to a single core on the central processing unit (CPU). The node shares the workspace with all instances, i.e. functions have to queue to be executed loosing LabVIEW's inherent ability for parallelization. Further, the data handling between both environments produces overhead, which slows down execution. MATLAB uses a just-in-time-compiling strategy, imposing an inherent handicap in terms of execution speed over compiled languages, such as C or LabVIEW. The 2018 32 bit release of LabVIEW was used because several extensions for hardware support were only available for this architecture. Unfortunately, MATLAB R2018b did not support this architecture anymore.

For all these reasons, selected functions were translated by the MATLAB-Coder to C code. As a peculiarity of the data transfer between LabVIEW and C / MATLAB, it must be noted that LabVIEW's memory management did not allow for dynamic memory change by an external source. The involved variables required manual wrappers in C code for the integration to LabVIEW. In a subsequent step, the source files were linked and compiled to a library, for which an optimized interface existed in LabVIEW. Such low-level C-based libraries often even outperform LabVIEW-written functions as was found out when benchmarking execution time. The procedure had the additional advantage that libraries could be compiled for different operating systems (e.g. DLLs in Windows and shared libraries in Linux) while the interface in LabVIEW remains the same.
The linker and compiler of the GNU compiler collection (GCC) were used, which was ported for Windows in minimalist GNU for Windows (MinGW). GCC itself is an open-source, cross-platform compiler collection for C and became the *de facto* standard in the free-software community. The suit is also recommended by The MathWorks for the Matlab-Coder Toolbox. MinGW / GCC was invoked manually including the dedicated wrappers for copying the data from the memory that LabVIEW had allocated.

Evaluation of the timing

There exist different strategies to integrate the MATLAB-based process model as an DLL in LabVIEW:

- as a single package calculating the force of an individual ensemble member based on the macroscopic engagement conditions,

- as a split function, first calculating the undeformed chip geometry and then simulating the individual force for each member of the ensemble, or
- by wrapping the entire forecasting in a single function.

The work refrained from implementing the process model in G again (LabVIEW's programming language), since it was originally developed, tested, and analyzed in MATLAB. While in the first two approaches, LabVIEW looped over the ensembles calling the DLL on each member, in the last approach, the looping was performed within the DLL. Integrating this looping minimized the number of external function calls but increased the maximum size of data, which was transferred between LabVIEW and the DLL. Tab. 6.4 lists the execution times on the CPU as an average of 500 simulations. Each entry represents the simulation of a single revolution with 237 samples and assuming an ensemble size of $J = 50$, i.e. the process model was called 11,850 times. The time per call of the process model is given in the parentheses. As a benchmark serves the timing in MATLAB. The number of function calls exhibited no influence. So neither did LabVIEW parallelize the loops, nor did handing over of data to an DLL slowed down the calculations.

Table 6.4: *Average CPU-times per simulation for different implementations of the ensemble forecast (average CPU-time per call of the process model)*

	single package	split function	ensemble
MATLAB	897 ms (75 µs)	595 ms (50 µs)	530 ms (45 µs)
DLL	712 ms (60 µs)	440 ms (37 µs)	440 ms (37 µs)

This short study fostered a deeper understanding on the computation at the interface of both programming languages, but the final question addresses the implementation of the entire EnKF. As before, different strategies for the implementation of the EnKF to LabVIEW could be followed:

- LabVIEW-based EnKF calling a MATLAB-based DLL for the ensemble forecast,
- pure MATLAB-based DLL, and
- hybrid: the calculations were performed in an DLL but the data storage (i.e. persistent variables) was organized by LabVIEW.

The last approach avoided initializing persistent variables and explicitly freeing their memory after the identification or experiment respectively.

The same conditions applied as before. This time, the parentheses in Tab. 6.5 included the average time per call of the EnKF ($J = 50$). The difference was marginal since the process model dominated the overall execution time of the filter. In the following, the purely DLL-based implementation was used because it favored a lean development circle.

Table 6.5: *Average CPU-times per simulation for different ways of how the EnKF and the process model are integrated to* LabVIEW *(average CPU-time per call of the EnKF)*

EnKF proc. mdl.	LabVIEW DLL	DLL	DLL+LabVIEW DLL
time	507 ms (2.14 ms)	466 ms (1.97 ms)	465 ms (1.96 ms)

6.5 Summary and findings on the model predictive force controller

MPC for force control in milling is much more than the pure control algorithm. This chapter explained the overall system (Fig. 6.1), how it was tuned, and implemented on the premise of direct feedback by the model identification. The discussions here focused on the most costly modules in terms of computation. New approaches were suggested, e.g. for the reference generator. The tuning of the controller was optimized by determining the minimum required prediction horizon N_2 and a sufficient resolution of the engagement conditions. Lastly, the implementation and integration to the MPFC system was studied, e.g. for the EnKF.

A major contribution to both, accuracy and transferability to other machine tools, was the development of a velocity filter on the basis of a classic KF. It used the commands of the NCU in conjunction with a model of each axis merging both with the measured velocity to provide a noise-free observation of the feed velocity. In this way, these measurement disturbances, which were filtered out by the lower control loop of the electric current, did not affect the MPFC system anymore. It further rendered the previous method of reading the feed velocity from the NCU via an external analog module (see [ADAM16a; STEM20]) no longer necessary.

Systematic design and tuning were necessary for a smooth integration of the novel instantaneous identification into the MPFC system. It fostered an in-depth understanding of the interaction between the individual blocks. The chapter states the configuration for the validation experiments in the next chapter (Chapter 7).

7 Validation of force control in milling

This chapter presents and discusses the results of the force control system in different configurations:

- MPFC with EnKF-based identification (uniformly / GAUSSIAN distributed initial ensemble, with / without ensemble inflation),
- MPFC with a repeated curve fit for identification as the state-of-the-art solution ([ADAM16a; STEM20]), and
- MPFC without identification, i.e. open-loop force control, closed-loop velocity control.

The first proves feasibility of the proposed system and examines different options for tuning. Strictly speaking, this system is the first closed-loop model predictive force control in milling. The remaining two versions state a benchmark for demonstrating the improvement over the state of the art. This reversed arc of suspense should help prioritize reading, presenting the most interesting results first, before discussing these findings in context of previous approaches. For maximum conciseness, the example geometry of Section 5.2.2 is manufactured.

7.1 MPFC with an EnKF-based identification

7.1.1 EnKF with uniform initial distribution

Fig. 7.1 presents the signals for the MPFC system as outlined in the previous chapter ($N_u = N_2 = 10$, $\boldsymbol{W_r} = 0.1\,\boldsymbol{I}$, $\boldsymbol{W}_u = 0.01\,\boldsymbol{I}$, $\boldsymbol{W_\xi} = 10^4$, $f_{t,max} =$ 0.25 mm) with an EnKF-based identification, which uses a uniformly distributed, random initial ensemble of size $J = 50$. The graphs are limited to the section of cutting. Special attention should be drawn to the top two rows: the active force F_a as the control variable and the feed velocity v_f as the manipulated variable. The coefficients of the process model and the engagement conditions complete the analysis. They are reflected in the reference of the feed velocity $v_{f,des}$ (black line). Crucial points are displayed enlarged. The position of these magnifications remains identical throughout this chapter to facilitate the comparison of the different system configurations.

The MPFC system operates the machine tool at the desired reference of the velocity and – curtailed slightly – of the active force. For clarification, the simulated maximum active force per revolution $F_{a,sim}|_{rev}^{max}$ is shown here. The simulation is based on the instantaneous feed and on the parameters

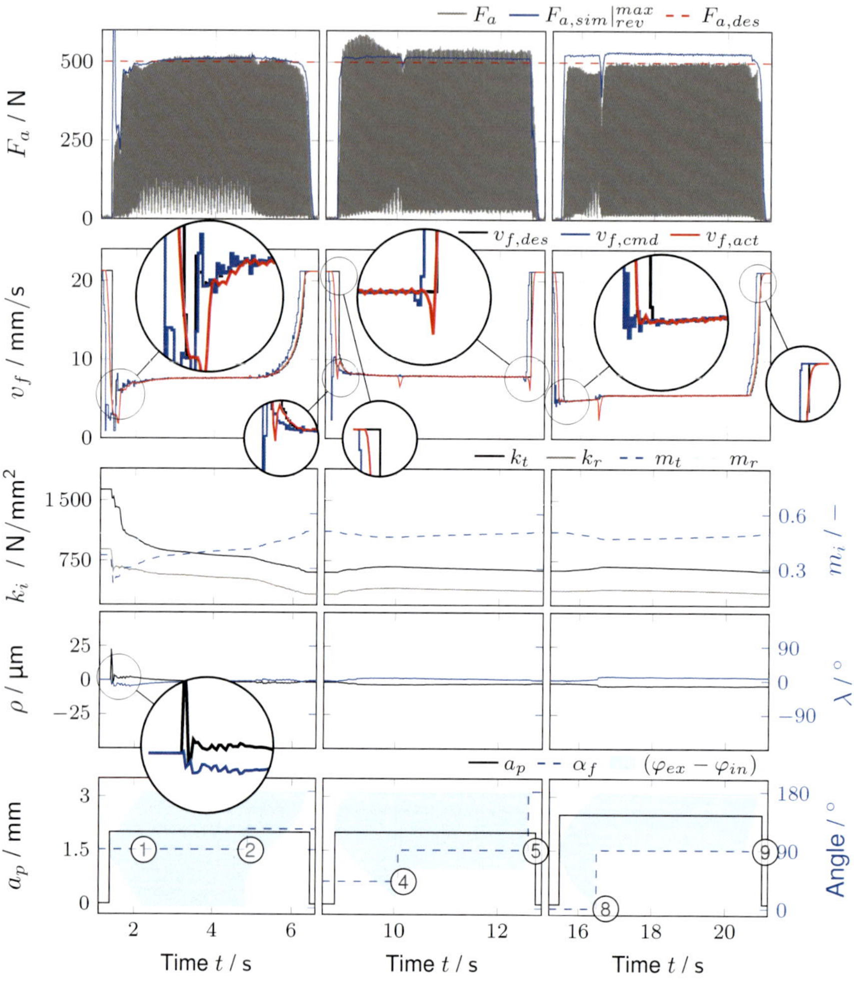

Figure 7.1: *MPC with EnKF ($J = 50$, uniformly distributed initial ensemble, S/N=15 dB assumed to determine $\boldsymbol{\Gamma}$); Fig. A.19 in Appendix A.3.4 illustrates the signals in position domain*

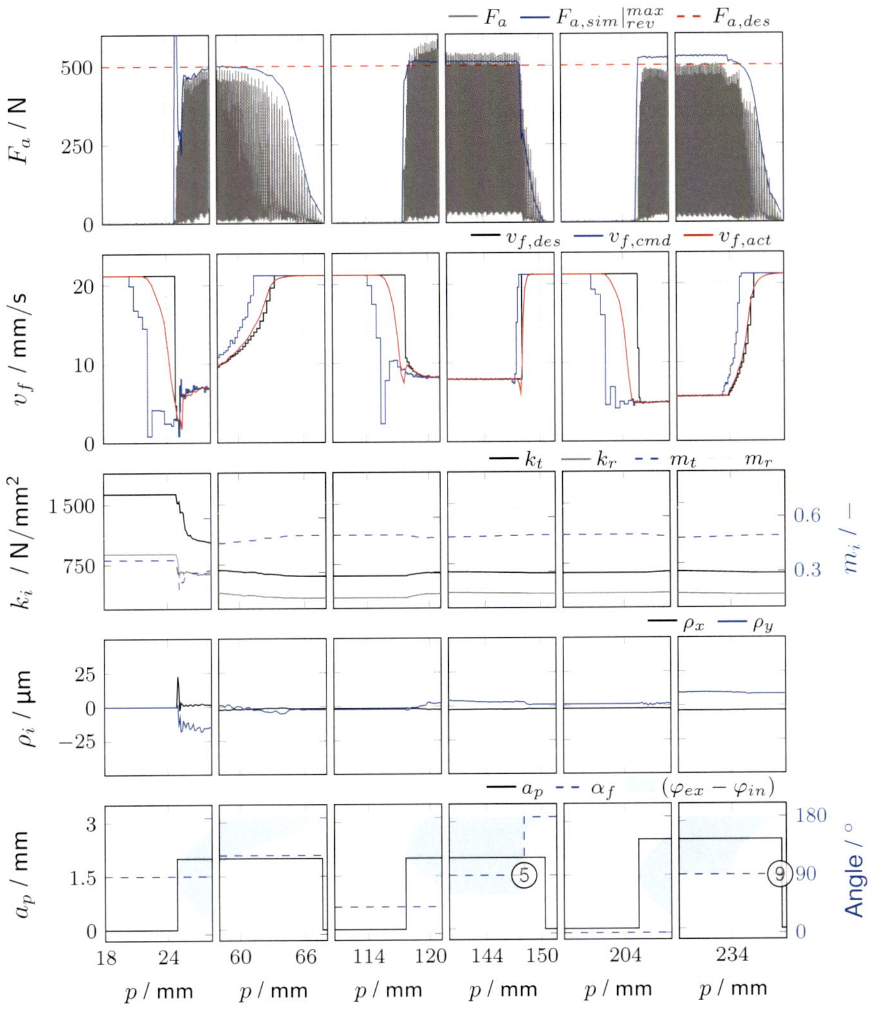

Figure 7.2: *Regions of changing depth of cut a_p of Fig. 7.1 but in position domain (MPC, EnKF with uniform $\boldsymbol{X}_0$)*

of the process model at the current time. This dependency becomes apparent at the very first cut ($t = 1.4\,\text{s}$), where the controller expected a force of $F_{a,sim}|_{rev}^{max} = 700\,\text{N}$ (the peak of the blue line in the top row, Fig. 7.1) reducing the feed accordingly. This inconsistent-looking behavior illustrates the effect of a wrong initial estimate of the parameters of the process model. The EnKF-based identification starts with a uniformly distributed, random ensemble from the technologically reasonable ranges derived before in Tab. 4.1 for the KIENZLE-model (Section 4.3) and for the KLINE & DEVOR-model in Section 4.4. The quick consolidation from this misalignment between simulation and measurements demonstrates the benefit of the novel identification method. The zooms on the radial deviation parameters ρ_i and on the feed velocity support this interpretation.
The constant offset between the simulated maximum force and the peak measurements in the third column are interesting to note. This error may originate from an incorrect identification of the radial deviation. Nonetheless, the MPC algorithm met the velocity reference.

One can see that the velocity command $v_{f,cmd}$ (blue line) goes down enough in advance of an abrupt upwards change in the reference $v_{f,des}$ (black line) so that the actual velocity $v_{f,act}$ of the tool is reduced just in time. The fourth zoom on the feed velocity v_f reveals a perhaps nonintuitive behavior: the controller commands the machine tool to be slower than necessary to keep its actual velocity as close as possible to the reference $\min(|v_{f,des} - v_{f,act}|)$. In fact, the objective is to maximize the position while respecting the velocity reference as a constraint. Maximizing the traveled distance at the end of the prediction horizon may provoke the seemingly contradictory behavior of a sudden drop in the actual velocity just before a vigorous acceleration. Here, the controller makes use of the characteristic that a step in velocity is reached independently of its height. Coming from a slower velocity allows a higher acceleration and an overall closer match to the desired reference. This behavior is only possible because the MPC algorithm maximizes the position over the prediction horizon. In a nutshell: the long-term benefit (maximum position) prevails over the short-term costs of a reduced speed.

Fig. 7.2 condenses the analysis to these regions where a change in depth of cut a_p occurs. The graphs are plotted in position domain. Each section is $10\,\text{mm}$ wide. Again, the plots to focus on are the ones of the feed velocity in the second row from the top. The figure proves how accurate the MPC algorithm works with regards to its reference ($v_{f,des}$, black line): it may not exceed it but minimizes the area to it. To be more precise, the position should be maximized, which translates into a minimal area between the actual velocity

$v_{f,act}$ and the reference $v_{f,des}$. The velocity reference also serves as an upper bound and must not be exceeded.

Appendix A.3.4 complements all figures presented in this chapter in position domain, i.e. along the tool path. The MPC algorithm solves the control problem along the tool path [SCHW17]. It presents a better illustration for comparing different control systems but also a less intuitive representation.

7.1.2 EnKF with uniform initial distribution and inflation

Inflating the ensemble on a regular basis improves its agility while maintaining the mean value as the current best guess. The larger the ensemble size, the less of the inflation is visible in the signals as an unsteady vibration. With an ensemble size of $J = 50$, the effect of the inflation is still visible in the signals, Fig. 7.3.

The hypothesis is that the inflated filter can follow time-variant parameters better, increasing overall accuracy. The repeated inflation holds the ensemble from a total collapse forcing it to a certain spread at distinct points in time. The minimum spread enables the filter to react to small and larger changes in the parameters. Fig. 7.3 confirms the previous findings on measurements of a process with constant feed (Section 5.2.3): how the parameters suggest to evolve smoother, e.g. there was no sudden change in the parameters of the radial deviation ρ_i as before (Fig. 7.1). However, this analysis is speculative because the experiments start from random initial ensembles.

It can be seen that both, the measured and the simulated maximum force, exceeded the desired maximum force $F_{a,des}$ slightly but almost constantly. The behavior became in particular apparent in Fig. 7.4 with enlarged regions. Nevertheless, MPC as velocity controller worked ideal. The velocity command $v_{f,cmd}$ was reduced just in time so that the actual velocity $v_{f,act}$ stayed below the reference $v_{f,des}$ but as close as possible to it. The drastic control effort was best visible in the steps down (first, third, and fifth column). An exemplary combination of an abrupt change and a continuous change in the reference gave the third column, where the tool entered the workpiece at an angle. The step change indicated the transition from aircut to cutting followed by a slow increase of the engagement angle resulting in this curved reference.

With regard to the uninflated EnKF-based MPFC, the ensemble inflation caused a smoother identification of the radial deviation parameters ρ_x and ρ_y. The presumed correct values were obtained at the beginning (about from time $t = 2.8\,\mathrm{s}$ on, Fig. 7.3) while Fig. 7.1 shows changes of the variables over the whole process. Fig. 7.5 displays the evolution of the coefficients of

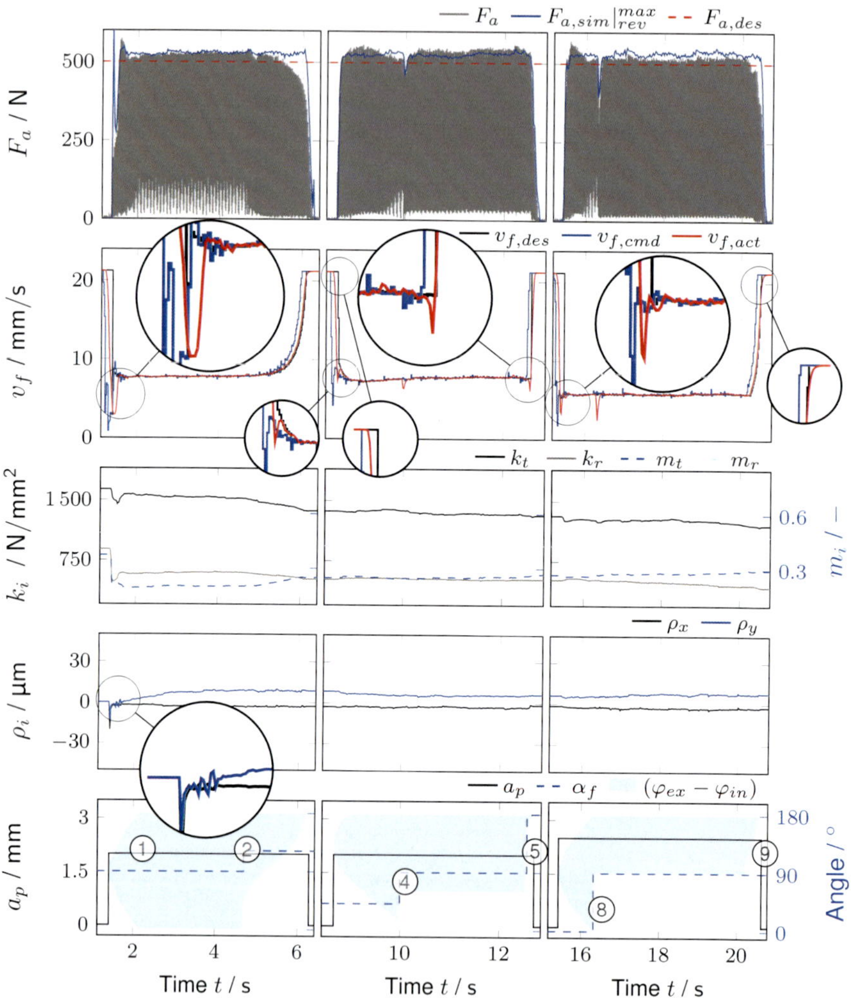

Figure 7.3: *MPC with EnKF ($J = 50$, uniformly distributed $\boldsymbol{X}_0$, ensemble inflation); Fig. A.20 in Appendix A.3.4 illustrates the signals in position domain*

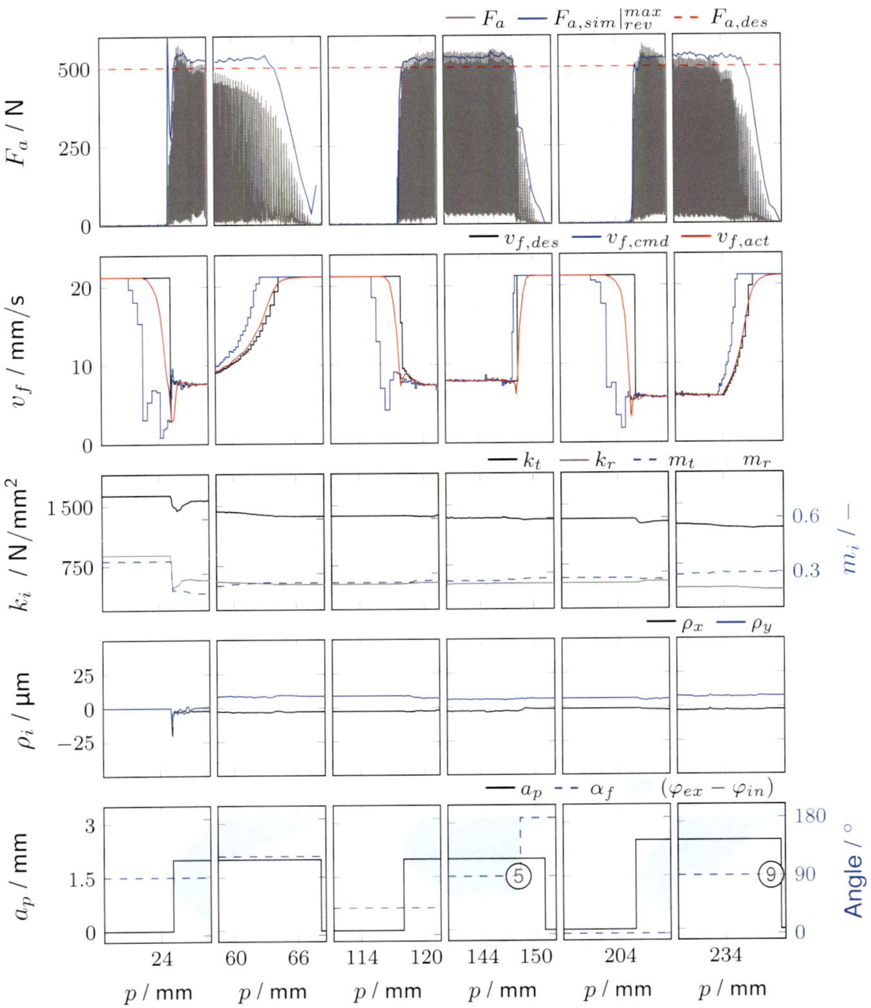

Figure 7.4: *Regions of changing depth of cut a_p of Fig. 7.3 but in position domain (MPC, EnKF with uniform $\boldsymbol{X}_0$, ensemble inflation)*

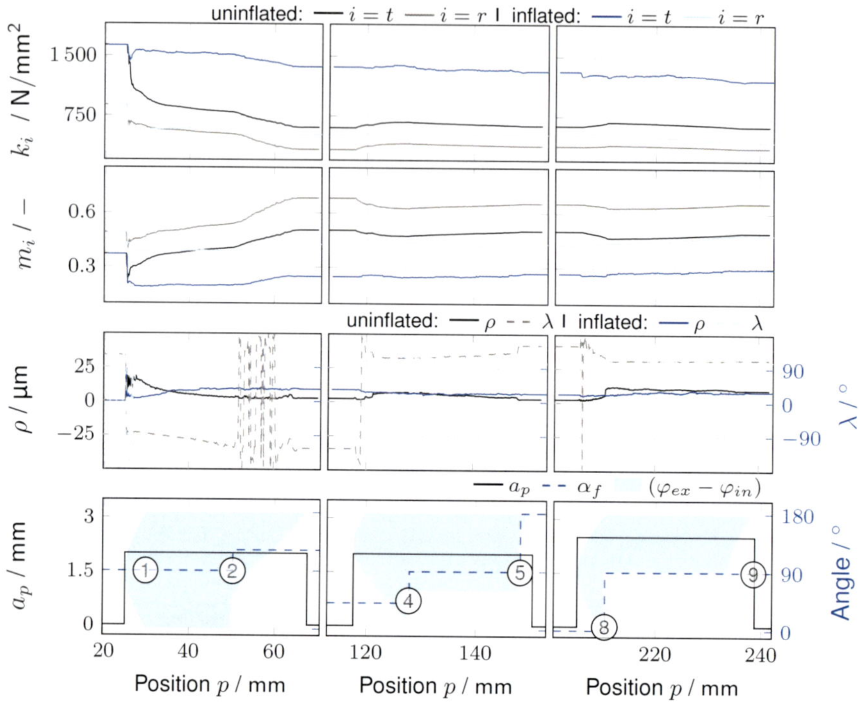

Figure 7.5: *Comparison of the coefficients of MPC with EnKF ($J = 50$, uniformly-distributed initial ensemble, S/N=15 dB assumed to determine Γ) and EnKF without and with ensemble inflation ($k^* = 50$, $\lambda^* = 10$). This compares Fig. 7.1 and Fig. 7.3 in position domain.*

both experiments on common axes. To restore technological interpretability of the parameters of the radial deviation model, they are provided in polar coordinates – although being identified as CARTESIAN parameters in both experiments. The frequent flipping at $\pm 180°$ emphasizes again the drawback of the periodic nature of polar coordinates. With regard to the comparison of both experiments, the radial deviation model stresses that the inflated EnKF (blue lines) leads to smoother identification to the presumably correct values.

An explanation for this steadier behavior is that the identification with ensemble inflation enables the EnKF to recover from local minima. This ability is especially important if states have a drastic influence on other states, such as the parameters of the radial deviation model have on the KIENZLE-model by affecting the undeformed chip thickness h. Fig. 7.5 illustrates that the KIENZLE-coefficients k_t , m_t and k_r , m_r change whenever the KLINE & DE-

VOR-parameters change. The argumentation must be considered with caution, as the parameters change synchronously and; therefore, there can be no order. However, the effect on the KIENZLE-coefficients seems to lasts longer.

7.1.3 EnKF with GAUSSIAN initial distribution

EVENSEN [EVEN03] recommended creating an ensemble by applying perturbations to a best-guess value to give the EnKF a good starting point. So the idea is to make use of the extensive literature on force model coefficients for the KIENZLE-model. A reasonable initial guess should reduce the sometimes drastic changes in the identified parameters at the first entrance of the tool into the material. Fig. 7.6 presents the signals in time domain while Fig. A.22 in Appendix A.3.4 zooms at the same selected regions as before in position domain, i.e. over the tool path.

The figures depict an even jumpier behavior at the first cut – although the mean of the initial ensemble is quite accurate as Fig. A.25 in Appendix A.3.4 demonstrates. Especially the radial offset ρ_i shows a unique behavior in the first region. It seems that the ensemble members were badly scattered due to their limited number ($J = 50$) and the GAUSSIAN distribution of them. This was especially problematic for the KLINE & DEVOR-model, whose search space apparently was not properly covered by a normal distribution around zero.
The idea relies on an accurate expectation value, i.e. the mean, and on a correct choice of the standard deviation. While the first is an obvious choice when literature on the considered material is available, a good standard deviation is much harder to determine. Using a standard deviation similar to the one of a uniform distribution ensures the same conditions as of the previous experiments.

Nonetheless, the measured force F_a and the simulated force $F_{a,sim}|_{rev}^{max}$ align well with the desired force in the rest of the first two sections. The last section exhibits a striking difference between the actual measurements and the perspective of the reference generator through $F_{a,sim}|_{rev}^{max}$ suggesting two findings:

1. A GAUSSIAN initial distribution reduces flexibility of the EnKF or its particles respectively.
2. The coefficients of the KIENZLE-force-model vary with the engagement conditions.

Engagement-dependent coefficients (k_i , m_i) were already suspected by previous works [ADAM16a; SCHW19a], but were partly attributed to the method of identification itself, which was a curve fit. In the previous configurations,

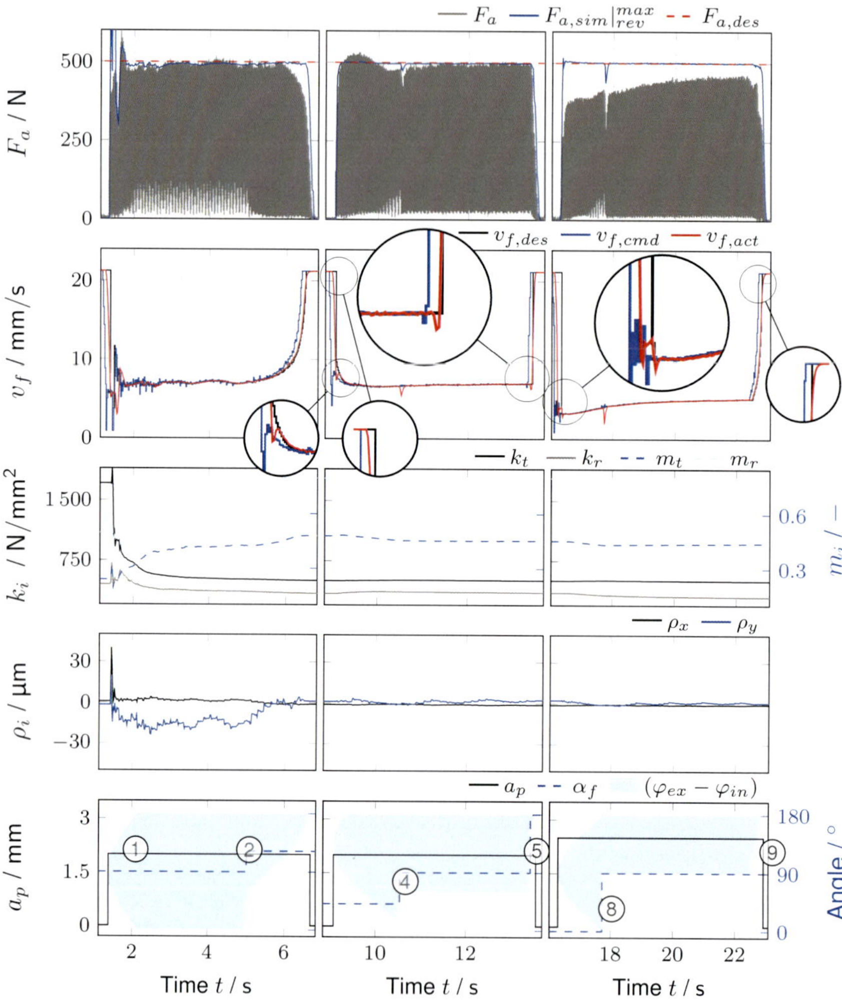

Figure 7.6: *MPC with EnKF ($J = 50$, GAUSSIAN distributed initial ensemble around $k_t = 1700\,\text{N/mm}^2$, $m_t = 0.18$, $k_r = 350\,\text{N/mm}^2$, $m_r = 0.55$, S/N=15 dB assumed to determine $\mathbf{\Gamma}$); Fig. A.21 in Appendix A.3.4 illustrates the signals in position domain*

the coefficients changed in the last section resulting in a quite constant force level. But only here it became obvious how wrong the force model may be if it was adjusted too slowly. The linear increase of the measured force in the third column was due to a gradual adjusting coefficient k_r , which decreased triggering a slow increase of the velocity reference $v_{f,des}$. The coefficients at the end of region two and at the beginning of the third region were identical. The fact that there was a drastic misfit in the force at the beginning of the third region but not in the second region suggested that the coefficients depended on the engagement conditions.

The massive change in all coefficients at the beginning of the cut indicates again that the KIENZLE-coefficients in literature may not accurately represent the force. This conclusion makes GAUSSIAN distributions especially critical because regions that are further away from the mean are underrepresented. Additionally, a GAUSSIAN distributed initial ensemble suffers from a lack of agility. This conclusion is drawn from the slow adjustment to the misfit of the force in the third region. The particles of the EnKF align fast, reducing the search space, giving up their agility. This problem may be faced by a larger ensemble size, a stronger variance inflation, or by repeated ensemble inflation with small values of λ^* for a larger blow-up.

7.2 Open-loop force control, closed-loop velocity control

The online identification of the force model closes the feedback loop of the force. Without it, the system reduces to an MPC-based velocity control. Its reference is determined by the engagement conditions along the tool path and by a constant force model. This construction is essentially an open-loop force control with an underlying (closed-loop) velocity control. The setup allows a detailed analysis of the controller itself eliminating the influence of a changing reference. It further demonstrates how the identification contributes to the control system.

Fig. 7.7 displays the force, the different velocities, and the engagement conditions for such an open-loop force control. The velocity reference was determined with the guess (i.e. the mean value) of a uniformly distributed initial ensemble ($k_t = 1762\,\text{N/mm}^2$, $m_t = 0.236$, $k_r = 447\,\text{N/mm}^2$, $m_r = 0.494$). The signals emphasized that the tuning of the MPC algorithm and the model of the machine tool were adequate. It further demonstrated that the data acquisition, the engagement simulation, and its synchronization worked correctly.

The control that Fig. 7.7 shows, relied on a reference $v_{f,des}$ calculated with the initial guess of an EnKF. A better choice was to consult literature for an adequate set of coefficients for the KIENZLE-model, e.g. [KÖNI82, p. 95]:

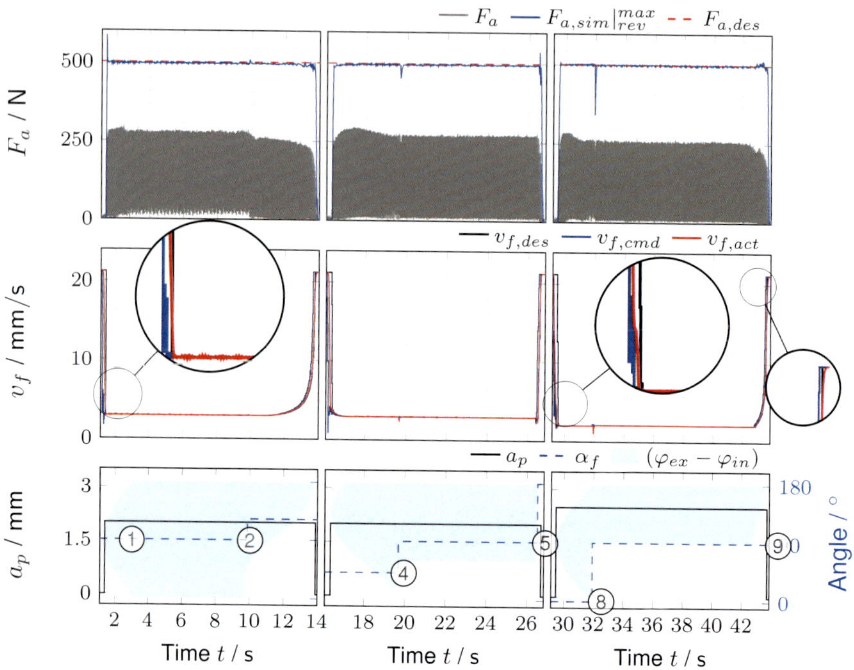

Figure 7.7: *MPC for velocity control. Velocity reference determined based on the initial value of an uniformly distributed ensemble (k_t = 1762 N/mm^2, m_t = 0.236, k_r = 447 N/mm^2, m_r = 0.494, no radial deviation). Fig. A.27 in Appendix A.3.4 illustrates the behavior in position domain.*

k_t = 1750 N/mm^2, $m_t = 0.18$, k_r = 350 N/mm^2, $m_r = 0.55$. The resulting process was much closer to the desired force, Appendix A.3.4 Fig. A.25. Since the models were not tuned for a specific tool, this open-loop force control did not consider any radial deviation of the tool.

The results already suggested a reasonably controlled process – in spite of no force feedback. The controller reliably met the reference or stayed below it. The open force-loop primarily resulted in a constant offset. While literature may provide good starting points, the accuracy deteriorates when the tool wears. To pronounce this effect, more workpieces must be manufactured provoking significant tool wear. Tool wear, as well as material-specific effects such as gradients in its hardness, might only be considered by a constant update of the reference trajectory.

There are two major differences to an *a priori* process optimization (e.g. by

VeriCut or SolidCAM) to the MPC-based velocity control: first, the design lacks feedback; second, they do not know the behavior of the machine tool. Therefore, preprogrammed feedrates do not react in time causing avoidable overshoots at steps down and an unnecessary slow reaction at upward steps.

7.3 MPFC with discrete identification

Applying a curve fit repeatedly is the current state of the art for online adaption of a mechanistic force model (cf. Section 2.2.5). A different, once examined approach is to use a RLS, see [ALTI17]. However, RLS has shown to be sensitive to measurement noise [SCHW19b].

The experiment here recreates the setup of Adams [ADAM16a]: a curve fit every 80 ms with the same initial values for the force model but the result of the previous fit as the initial guess for the parameters of the radial deviation model. The result is smoothed exponentially with a factor $\alpha_{smth} = 0.43$ [ADAM16a, p. 108]. Fig. 7.8 presents the results. The notable difference to the previous work is that it does not use a cross-correlation before the fit, with which Adams [ADAM16a, pp. 55–64] determined the position of the cutting edges in the force measurements. Since the position is measured directly, it renders the cross-correlation obsolete.

The erratic velocity reference at the first step downwards (first column), showing a bump to 10 mm/s, was caused by the start of the identification. Abrupt changes in the process model and, with it, in the velocity reference were the reason why Adams [ADAM16a] smoothed the identified parameters exponentially. It was especially volatile at small forces and potentially devastating when the tool entered the material. To exclude these conditions, the discrete identification was limited to a minimum engagement angle and a minimum value of the mean force. The presented experiment refrained from applying multiple thresholds and used a minimum mean active force $\overline{\boldsymbol{F}_a} = 70$ N as exclusive limit. The stepwise changes in the model coefficients was visible in the graphs, clearly indicating the effect of the additional smoothing. For example, at the second entry of the tool (beginning of second region), the gain coefficient k_t showed an inverse exponential progression. Moreover, the MPFC system was ported to a different machine tool than in [ADAM16a] and used the same system settings as described in Chapter 6. In this way, the only difference to the results above was the identification method.

The system constantly underestimated the force resulting in a too high force when calculating the maximum active force $F_{a,sim}|_{rev}^{max}$ from the velocity reference $v_{f,des}$. It was noticeable that the coefficient k_t was much higher than in

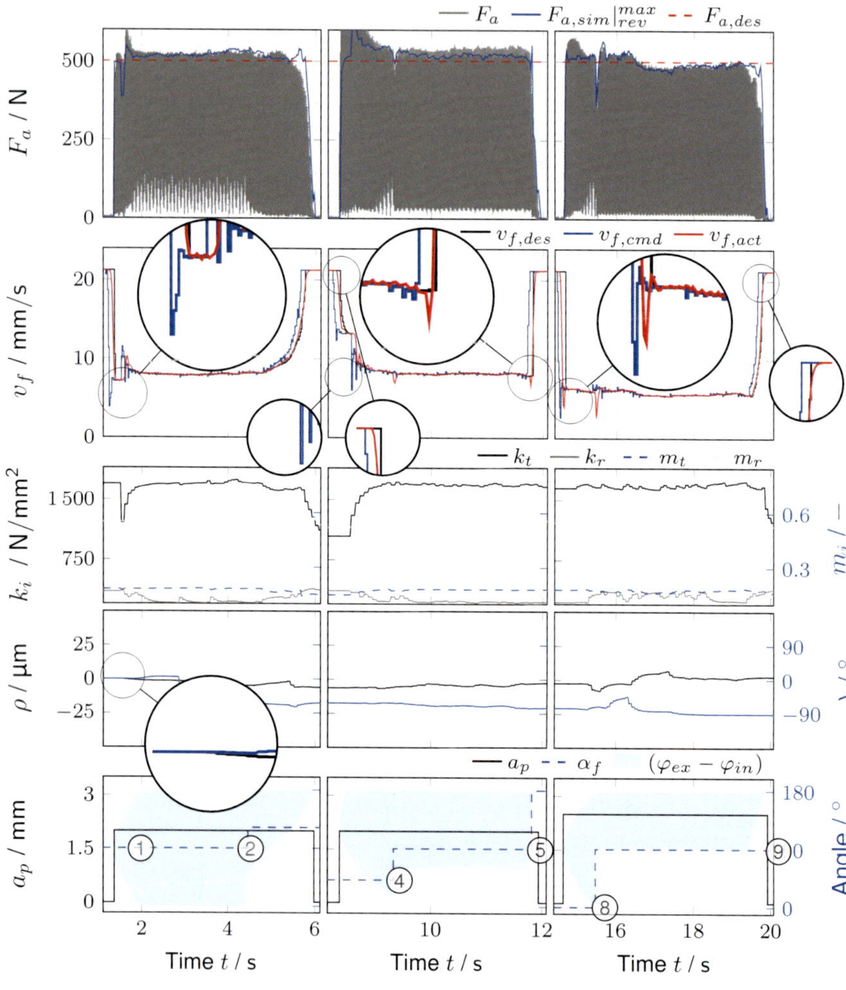

Figure 7.8: *MPFC with a curve fit for model identification in accordance to [ADAM16a; SCHW19a; STEM20]; Fig. A.23 in Appendix A.3.4 illustrates the signals in position domain*

other experiments. Further, the radial deviation coefficients constantly grew. These observations suggested a limiting suitability of the curve-fit-based identification for online control. ADAMS [ADAM16a, p. 60] was obliged to restrict the identification to a small range of the coefficients ($10 - 40\,\%$ of the range used in this work here, Tab. 4.1). This restrictive boundaries required an accurate guess of the range sacrificing the demand for general applicability and usability.

7.4 Summary and findings of EnKF-based MPFC in milling

The extensive evaluation of the MPFC system demonstrated the contribution of this work. The comparison with the existing concept was particularly revealing, pointing out the benefit of a recursive identification – which was continuous feedback from the perspective of the MPC algorithm. The curve-fit-based identification resembled a cascaded control with a higher-level, low-frequency force controller (80-400 ms due to smoothing) and an underlying MPC-based velocity controller at 20 ms.
Nevertheless, the active force did not always match the force reference when controlled with an EnKF-based identification. Violations of the desired force reference $F_{a,des}$ may be attributed to inaccuracies of the process model. One possible cause was a missing low-pass filter in the model of the EnKF. The lack of signal conditioning induced a systematic error, which became larger the larger the undeformed chip thickness was when the cutting edge entered the material. The corresponding signal impulse was largest at the last section, where half immersion milling occurred $\varphi_{in} = 90°$.

Tab. 7.1 presents a different viewpoint by shrinking the comparison of all configurations to process times: the time the numerical controller (NC)-program took, the time between the first cut and the last cut T_{prc}, and the time that the cutter was engaged T_{cut}. Two processes with constant feed are presented for comparison. One with the same cutting velocity ($v_c = 80\,\mathrm{m/min}$) as the experiments with MPFC, and one with a smaller cutting velocity and twice the feed ($v_c = 70\,\mathrm{m/min}$, $f_t = 0.1\,\mathrm{mm}$). Both processes lie within the recommended design space of the manufacturer. The first produces a force constantly below the desired force reference $F_{a,des}$, while the second exceeds the desired force in the third column, where the depth of cut increases. This design was used in Chapter 5 to test the EnKF.

In Tab. 7.1, one can see the reduction of the cutting time T_{cut} for controlled processes – if the model was correct. This reduction presented the pure gain in productivity because positioning movements in the air could also be optimized manually. The open-loop force controller with the initial guess of an

EnKF-based process, i.e. the initial guess of a uniformly distributed random ensemble, was much slower than the conventionally designed process. This disadvantage could be compensated with a good initial (constant) force model. However, an EnKF with a GAUSSIAN initial ensemble distributed around an accurate starting point failed – most likely because of an unfavorable choice of the standard deviation and, in this consequence, that the small ensemble ($J = 50$) aligned too fast on an incorrect value. The times in Tab. 7.1 quantify this unexpected result.
Moreover, the numbers shed a light on a potential dilemma of the MPFC: it was able to control the force independently of the wear state of the tool. But this ability did not always result in a faster, more productive process. Worn tools might provoke a significant higher force resulting in a slower-than-intended process. Nevertheless, the process would still be safe with regard to tool failure and with regard to the deviation of the tool, i.e. of meeting the required dimensional accuracy. Only a holistic evaluation could answer the question up to what point the MPFC provides an economic benefit.

Table 7.1: *Comparison of the total time T_{all}, the time from the first to the last cut T_{prc}, and the cutting time T_{cut}*

Configuration	T_{all}	T_{prc}	T_{cut}
uncontrolled $v_c = 80\,\mathrm{m/min}$, $f_t = 0.05\,\mathrm{mm}$	64.14 s	50.70 s	22.78 s
uncontrolled $v_c = 70\,\mathrm{m/min}$, $f_t = 0.1\,\mathrm{mm}$	36.33 s	28.64 s	14.00 s
open, $\overline{\boldsymbol{X}_0}$	45.36 s	42.38 s	17.07 s
open, [KÖNI82, p. 95]	23.11 s	20.20 s	8.48 s
fit every 80 ms	21.60 s	18.56 s	7.70 s
EnKF, uniform $\boldsymbol{X}_0$	22.57 s	19.64 s	8.14 s
EnKF, uniform $\boldsymbol{X}_0$, $k^* = 50$, $\lambda^* = 10$	22.35 s	19.24 s	8.10 s
EnKF, GAUSSIAN $\boldsymbol{X}_0$	24.59 s	22.94 s	8.92 s
EnKF, uniform $\boldsymbol{X}_0$, $f_{t,max} = 0.39\,\mathrm{mm}$	19.85 s	17.74 s	7.26 s

The control system enhanced the non-cutting time even more if the maximum allowed feed per tooth was further increased. Appendix A.3.4 presents Fig. A.28 and Fig. A.29, in which the upper limit was increased by 40-60 % to $f_{t,max} = 0.39\,\mathrm{mm}$ ($v_{f,max} = 2000\,\mathrm{mm/min}$). Tab. 7.1 depicts the corresponding times, confirming that increasing the upper limit of the feed velocity primarily reduced the total time T_{all} and the process time T_{prc} by faster movements in the air.

8 Conclusion

8.1 Summary and discussion

The motivation of this thesis was to control the maximum active force $F_{a,max}$ per revolution in a milling process. For this, model-based predictive control (MPC) was already introduced before by STEMMLER ET AL. [STEM16] and coworkers. They built an model predictive force controller (MPFC) using two separate models, one of the machine tool and one of the process. The MPC-based controller used the model of the machine tool in an optimization at every time step implicitly deriving the current control law. The reference was determined with the help of the process model. It closes the feedback loop of the force by continuously updating or "identifying" the process model.
The state-of-the-art approach for online model identification relied on repeated, uncorrelated curve fits at intervals of 80-200 ms, while the MPC-based controller ran at a sampling time of $T_s = 20\,\text{ms}$. Due to the discontinuity in the feedback induced by uncorrelated models, the MPFC required additional smoothing of the identified parameters at sudden changes in order to maintain a stable control behavior. Within these intervals, the system represented an open-loop force controller with underlying closed-loop velocity control based on MPC. The smoothing delays the feedback and, thus, additionally enlarges the intervals of open-loop control. The thesis in hand claimed that the identification can be done by a recursive method providing a continuous feedback at every time instant. For such an instantaneous identification, the particle-based method ensemble KALMAN filter (EnKF) was introduced. The method stated the major progress to previous works (cf. [ADAM16a; STEM20]).

The EnKF recursively estimates model parameters at every time step automatically providing a continuous calibration of the model. The ensemble members are particles approximating the posterior (probability) distribution of the model parameters. Hence, more particles provide a better approximation. The expectancy value is the mean value of all particles (aka. *ensemble members*) and serves as the current best guess of the EnKF at every time instant. The size of the filter, in terms of number of particles, has a cubic relation to convergence rate. Studying the optimal size suggested that an ensemble size within the range $J \in [25, 100]$ was sufficient for the identification of a process model with eight free parameters. The process model was created from the well-established mechanistic force model according to KIENZLE [KIEN52] in conjunction with the popular KLINE & DEVOR-model [KLIN83] to model the radial deviation of the tool. For identification, the parameters of the radial offset were expressed in CARTESIAN coordinates rather than in the usual polar coordinates. This transformation was necessary because the EnKF required a mathematically continuous definition of the mean. The periodic nature of polar coordinates prohibited a continuous function of the mean.

In order to determine the undeformed chip thickness h, which was required for the process model, the current engagement conditions and the position of the cutting edge(s) needed to be known. The radial deviation model adjusts the undeformed chip thickness for the force model. The models were calibrated recursively on each instantaneous force measurement. Studying robustness towards measurement noise demonstrated that the EnKF was remarkably insensitive to it. The estimation of the EnKF could be enhanced further by fusing the partly redundant sensor signals of the cutting torque at the spindle with the CARTESIAN force vector of the table dynamometer. The idea was that different sensors potentially provide different signal quality and, thus, a fusion my enhance the overall parameter estimation of the process model. Simulation demonstrated the correctness of this assumption.

To account for time-variant model parameters, e.g. due to tool wear, a repeated ensemble inflation was suggested. Preliminary results from simulations (cf. [SCHW20]) were confirmed on real measurements: small but frequent inflation was to be preferred over large but rare inflation. The study on real measurements revealed the modeled force missed the effect of a low-pass filter, which led to a comparably large error at rising force signals (in down milling). In particular at down-milling, the entry of a cutting edge caused a seep signal slope (if it was not slot milling). The low-pass filter smoothed this slope suggesting a large error to the identification forcing the EnKF to adjust the model parameters. The EnKF was only capable of doing this adaption because it maintained its agility through ensemble inflation – this became obvious in the last region of the example workpiece, where half-immersion cutting $a_e = 50\,\%$ was conducted. It is to be answered whether the a_e-dependency of the parameters of the KIENZLE-model were an artifact caused by the missing low-pass filter. Therefore, the integration of a low-pass filter must be subject of future research.

For the novel online identification, the MPFC of previous works was revised to meet the real-time requirements of the individual system modules. Along the way, the MPFC was put on a proper basis with regard to its design and tuning, e.g. by deriving the length of the prediction horizon based on the machine tool model.

The systematic evaluation of different configurations of the MPFC system demonstrated its feasibility, flexibility, and the potential economical benefit of an EnKF-MPFC. The advantage over the state-of-the-research approaches

was first and foremost robustness. The EnKF provided an instantaneous and recursive identification, which resembled continuous feedback for the controller. The discrete changes in the previously used curve-fit-based identification caused erratic control behavior and severe violations of the force constraints. (However, exceeding the force reference and, with it, the velocity bounds inevitably resulted in a faster, more productive process – at the cost of its safety.)
The advantage over the state-of-the-industry solutions was a reduction in the manufacturing time by 40-60 % compared to a conventionally designed process with constant feed. This consideration did not yet include a better utilization of the tool life as a result of a more homogeneous load on the tool. Nor did it include the potential benefit of reduced scrap and downtimes by limiting of the deflection of the tool through an upper limit of the active force.

8.2 Outlook

Although entering a mature stage, model predictive force control for milling still holds a broad bucket of questions for production engineering. The most recent is to combine it with works on a current-based force measurement, see [MIUR18; MIUR19], to substitute the table dynamometer as force sensor. These indirect measurements exhibit larger noise levels at the slowly moving linear axes and a better quality for the fast-rotating spindle. Combining this partly redundant information may enhance the overall estimate. Section 5.1.2 demonstrated that the just-introduced EnKF is capable of doing this fusion. Establishing a permanently closed MPFC presents the starting point for an application-driven development, which leads to building technologically motivated force references, considering arbitrary tool geometries, and an economical evaluation of the system. From a technological perspective, the extension to five-axis machining is especially challenging. The rotary axes introduce noncontinuous behavior and simultaneous movement in space imposes a special challenge on the measurement of the force.

The question of a nonlinear MPC lies on the way towards industrial application, five-axis machining, and higher robustness. Integrating the process model in the MPFC makes the system generic again, where the combination of a reference generator and an MPC for velocity control is after all still a workaround. The system offers a multitude of fields for further research in different perspectives:

- Production engineering
  - substituting the table dynamometer: current-based force estimation and sensor fusion,

  - extension to five-axis machining,
  - extension to ball-shaped tools and finish milling,
  - deriving a quality-based force reference from the tolerances of the workpiece,
  - considering different costs, e.g. the cutting force or creating a multi-objective optimization, or
  - examining the effect on tool life and quantifying the overall economical benefit.
- Control engineering
  - integrating the process model to the MPC, rendering it nonlinear,
  - using a nonlinear machine tool model to account for the additional feedforward control of the NC.
- Computer science
  - transfer to real-time hardware in order to obtain a deterministic system,
  - distributed computing in a low-latency network.

Moreover, the work on an online identification provided a new twist for a different field of research in cutting technology: since the radial deviation affects the shape of stability lobe diagrams [ZHAN18a], knowing the radial offset allows to individualize these diagrams [SCHM07]. So an in-process identification can contribute to a stable milling process.

Fazit

Zusammenfassung und Diskussion

Das Ziel der Arbeit war die Regelung des Maximums der Aktivkraft $F_{a,max}$ pro Werkzeugumdrehung beim Fräsen. Dafür wurde zuvor bereits ein modellbasierter, prädiktive Regler (MPC) eingeführt, welcher anhand eines Modells der Maschine und des Prozesses in jedem Zeitschritt ein Optimierungsproblem löst und so das aktuelle Reglergesetz implizit bestimmt. Der Kraftregelkreis wird über das Anpassen bzw. „Identifizieren“ des Prozessmodells geschlossen. Der Stand der Technik stützte sich auf wiederholte Optimierungsverfahren in Intervallen alle $80\,\mathrm{ms}$ bis $200\,\mathrm{ms}$. Die so identifizierten Prozessmodelle waren unabhängig voneinander und die abrupten Übergänge zwischen ihnen mussten geglättet werden, um ein stabiles Verhalten des Reglers zu erreichen. Innerhalb dieser Intervalle stellte das System eine Kraftsteuerung (offener Kraftregelkreis) mit unterlagerter, MPC-basierter Geschwindigkeitsregelung dar (geschlossener Geschwindigkeitsregelkreis). Durch die Glättung wurde die Rückführung des Kraftregelkreises zusätzlich verzögert. Die Hypothese der vorliegenden Arbeit war, dass dies mit Hilfe einer rekursiven Identifikation geschlossen werden könne. Dazu wurde ein stochastischer Zustandsschätzer, ein Partikelfilter eingeführt, welches den wesentlichen Fortschritt zu bestehenden Arbeiten darstellt (vgl. [ADAM16a; STEM20]).

Die Arbeit schließt den Kraftregelkreis durch ein Ensemble KALMAN Filter (EnKF). Dieses schätzt die Modellparameter an jedem neuen Messpunkt rekursiv und ermöglicht damit eine kontinuierliche Anpassung des Prozessmodells. Die Partikel des EnKF approximieren die *posteriori* Wahrscheinlichkeitsverteilung der Modellparameter. Je größer das Ensemble, d.h. je mehr Partikel dieses beinhaltet, desto genauer ist diese Näherung. Der Erwartungswert ist der Mittelwert aller Mitglieder des Ensembles und stellt immer den aktuellen Schätzwert des Filters dar. Das Prozessmodell besteht aus dem mechanistischen Kraftmodell nach KIENZLE [KIEN52] und einem Modell für den Rundlauffehler eines Fräswerkzeugs. Dafür wurde das weitverbreitete Modell nach KLINE und DEVOR [KLIN83] verwendet, dessen Parameter jedoch für die Identifikation mittels EnKF als kartesische Koordinaten anstatt den üblichen Polarkoordinaten ausgedrückt wurde. Dies war nötig, da das EnKF eine mathematisch stetige Funktion für den Mittelwert benötigt, was wegen der Periodizität von Polarkoordinaten in diesen nicht möglich war. Systematische Untersuchungen hinsichtlich der Robustheit der neuen Identifikationsmethode ließen auf eine ausgeprägte Unempfindlichkeit gegenüber Messrauschen schließen. Ferner konnte gezeigt werden, dass eine Ensemblegröße im Bereich $J \in [25, 100]$ für die Identifikation des Prozessmodells mit acht freien Parametern ausreichend war.

Für die Identifikation in einem Abtastschritt sind die aktuellen Eingriffsbedingungen und die Position der Schneidkante(n) notwendig, um die Spanungsdicke h zu jedem Zeitpunkt zu bestimmen. Darüber hinaus können durch das EnKF Messwerte von verschiedenen Kraftsensoren mit potentiell unterschiedlicher Signalqualität fusioniert werden, um die Identifikation zu verbessern. Zur Berücksichtigung zeitlich veränderlicher Modellparameter, bspw. durch fortschreitenden Werkzeugverschleiß, wurde vorgeschlagen, das Ensemble in regelmäßigen Abständen um seinen aktuellen Mittelwert herum zu vergrößern (*ensemble inflation*). Vorherige, auf Simulationen basierende Ergebnisse (vgl. [SCHW20]) konnten experimentell bestätigt werden: kleines, aber häufiges Aufpumpen ist seltenem, großem vorzuziehen. Die Untersuchung realer Messsignale offenbarte, dass die fehlende Modellierung des Tiefpassfilters des Kraftsensors zu einem relativ großen Fehler führen kann, insbesondere bei abrupten Änderungen der Kraftsignale. Daher muss die Integration eines Tiefpassfilters im Prozessmodell Gegenstand zukünftiger Weiterentwicklungen sein.

Für die Integration des neuen, „instantanen“ Identifikationsverfahrens wurde das bestehende Konzept der MPC für das Fräsen vollständig hinsichtlich der Auswirkungen auf die Echtzeitanforderungen der unterschiedlichen Module überarbeitet. Im Zuge dessen wurde das Gesamtsystem systematisch ausgelegt, z.B. wurde ein Vorgehen beschrieben, um die notwendige Länge des Prädiktionshorizonts der MPC anhand des Modells der Werkzeugmaschine zu bestimmen.

Die systematische Auswertung unterschiedlicher Konfigurationen des MPC-Systems demonstrierte die Machbarkeit, Flexibilität und den potentiellen wirtschaftlichen Nutzen einer MPC mit EnKF-basierter Rückführung. Die Verbesserung gegenüber dem Stand der Forschung ist vor allem eine höhere Robustheit. Das EnKF ermöglicht eine kontinuierliche – für den Regler sofortige – Rückführung der Kraft bzw. des Kraftmodells. Die bisher verwendeten diskreten Identifikationen anhand von Optimierungsverfahren hatten teils signifikante Verletzungen der gewünschten Kraftobergrenze zur Folge. (Allerdings führte das Überschreiten des Kraftsollwertes und damit der Sollgeschwindigkeit zwangsläufig zu einem schnelleren, produktiveren Prozess – jedoch auf Kosten seiner Sicherheit.)
Gegenüber dem Stand der Technik konnte die Fertigungszeit um 40-60 % reduziert werden, verglichen mit einem konventionell ausgelegten Prozess mit konstantem Vorschub. Diese Betrachtung beinhaltete noch keine mögliche bessere Ausnutzung der Werkzeugstandzeit durch eine homogenere

Belastung des Werkzeugs, oder einer Reduktion von Stillstandszeiten und von Ausschuss durch die Begrenzung der Werkzeugdurchbiegung anhand einer Begrenzung der Aktivkraft.

Ausblick

Trotz langjähriger Forschung an einer MPC zur Kraftregelung beim Fräsen wirft ein solches System noch immer vielfältige Fragen für die Produktionstechnik auf, die es zu erforschen gilt. Eine dieser ist die Frage, wie das MPC-System mit den Arbeiten zu einer indirekten, strombasierten Kraftmessung (siehe [MIUR18; MIUR19]) kombiniert werden kann, um die piezoelektrische Kraftmessplattform als Kraftsensor zu ersetzen. Diese indirekten Kraftmessungen weisen ein höheres Signalrauschen an den sich verhältnismäßig langsam bewegenden Linearachsen auf. Eine indirekte Messung des Drehmoments über die sich schnell drehende Spindel weist hingegen eine höhere Signalqualität auf. Das EnKF ist in der Lage, diese teilweise redundanten Informationen zu kombinieren und so die Gesamtschätzung zu verbessern, vgl. Abschnitt 5.1.2. Eine anwendergetriebene Forschung würde ebenfalls die Ableitung technologisch-motivierter Kraftreferenzen, bspw. über die Modellierung der Werkzeugdurchbiegung, behandeln und die Erweiterung des verwendeten Prozessmodells hin zu beliebigen Werkzeuggeometrien, sowie das System ganzheitlich hinsichtlich seiner Wirtschaftlichkeit bewerten. Aus technologischer Sicht stellt eine Erweiterung auf das fünfachsige Fräsen eine besondere Herausforderung dar. Die Berücksichtigung der rotatorischen Antriebsachsen führt zu einem nicht-stetigen Maschinenmodell und die simultane Bewegung der Achsen im Raum erschwert die Messung der Kraft.

Hinsichtlich der Regelungstechnik stellt sich die Frage nach einer nichtlinearen MPC – insbesondere vor dem Hintergrund des fünfachsigen Fräsens. Momentan handelt es sich bei der MPC um einen Geschwindigkeitsregler, welcher durch die Invertierung des Prozessmodells im Referenzgenerator zu einem Kraftregler wird. Die Integration des Prozessmodells in die MPC würde eine nichtlineare Formulierung erfordern, das System jedoch auch generisch machen. Die verschiedenen Forschungsfragen sind im Folgenden, nach Forschungsdisziplinen getrennt, stichpunktartig aufgelistet:

- Produktionstechnik
  - Ersetzen der Kraftmessplattform durch eine strombasierte Kraftmessung und Sensorfusion
  - Erweitern zum fünfachsigen Fräsen
  - Erweitern für Kugelkopffräsern und auf das Schlichtfräsen

  - Ableiten und nutzen einer Kraftreferenz anhand der Maßhaltigkeitsanforderungen des Bauteils
  - Untersuchen anderer Regelgrößen, z.B. die Schnittkraft oder eine Mehrgrößenoptimierung
  - Bewerten der Auswirkungen auf den Standweg des Werkzeugs und auf die Wirtschaftlichkeit
- Regelungstechnik
  - Integrieren des Prozessmodells in die (dann nichtlineare) MPC
  - Berücksichtigen der Vorsteuerung der NC in einem nichtlinearen Modell des Maschinenverhaltens
- Computerwissenschaften
  - Implementierung auf einem Echtzeitsystem (Hardware) für eine vollständig deterministische Regelung
  - verteilen der Berechnungen in einem *low-latency* Netzwerk

Darüber hinaus ermöglicht die Arbeit an einer Online-Identifikation eine neue Sicht auf andere Forschungsgebiete in der Zerspanung: ZHANG u. a. [ZHAN18a] zeigte, dass der Rundlauffehler die Form einer Stabilitätskarte beeinflusst. Daher kann durch eine Identifikation im Prozess eine individuelle Stabilitätskarte für jedes Werkzeug erstellt werden [SCHM07] und trägt so zu einem stabilen Fräsprozess bei.

Appendix

A.1 Notes on the ensemble KALMAN filter

A.1.1 Matrix dimensions of the EnKF

The ensemble matrix $\boldsymbol{X}$ (dimension $(N_{\boldsymbol{x}} \times J)$) consists of J state vectors, each with $N_{\boldsymbol{x}}$ states. The number of states that can be observed N_{obs} is smaller or equal to the number of states $N_{\text{obs}} \leq N_{\boldsymbol{x}}$. The observation matrix $\boldsymbol{H}$ has $(N_{\text{obs}} \times N_{\boldsymbol{x}})$ dimensions. The different matrix calculations are illustrated below.

$$\underbrace{\boldsymbol{X}}_{(N_{\boldsymbol{x}} \times J)} = \left[\underbrace{\boldsymbol{x}^1}_{(N_{\boldsymbol{x}} \times 1)}, \underbrace{\boldsymbol{x}^2}_{(N_{\boldsymbol{x}} \times 1)}, \ldots, \underbrace{\boldsymbol{x}^J}_{(N_{\boldsymbol{x}} \times 1)} \right] \tag{A.1}$$

$$\underbrace{\overline{\boldsymbol{P}}}_{(N_{\boldsymbol{x}} \times N_{\boldsymbol{x}})} = \frac{1}{J} \underbrace{\boldsymbol{X}}_{(N_{\boldsymbol{x}} \times J)} \underbrace{\boldsymbol{X}^{\mathsf{T}}}_{(J \times N_{\boldsymbol{x}})} \tag{A.2}$$

$$\underbrace{\boldsymbol{z}^j}_{(N_{\text{obs}} \times 1)} = \underbrace{\boldsymbol{z}}_{(N_{\text{obs}} \times 1)} + \underbrace{\boldsymbol{\eta}^j}_{(N_{\text{obs}} \times 1)} \tag{A.3}$$

$$\Rightarrow \underbrace{\overline{\boldsymbol{R}}}_{(N_{\text{obs}} \times N_{\text{obs}})} = \frac{1}{J} \underbrace{\boldsymbol{\eta}}_{(N_{\text{obs}} \times J)} \underbrace{\boldsymbol{\eta}^{\mathsf{T}}}_{(J \times N_{\text{obs}})} \tag{A.4}$$

$$\underbrace{y^j}_{(N_{\text{obs}} \times 1)} = \underbrace{\boldsymbol{H}}_{(N_{\text{obs}} \times N_{\boldsymbol{x}})} \underbrace{\boldsymbol{x}^j}_{(N_{\boldsymbol{x}} \times 1)} \tag{A.5}$$

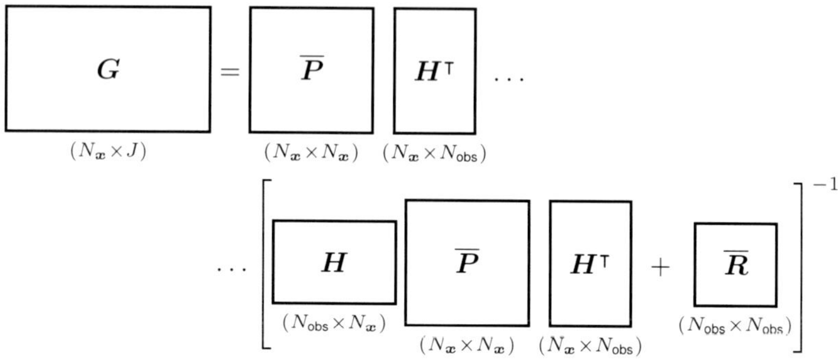

(A.6)

A.1.2 Alternative formulation of the EnKF

In the discipline of analyzing the well-posedness, stability, and the general behavior of EnKF, a different formulation is used than in Section 4.2.1. It moves away from the "engineering" perspective, which is motivated from the classic KF. Both formulations are identical in theory but may differ on hardware due to numerical rounding. The matrices $\boldsymbol{Z}$ and $\hat{\boldsymbol{Y}}$ denote the matrices build from the augmented measurements $\boldsymbol{z}^j$ for every ensemble and the predicted values y^j respectively (both have the dimension $(m \times J)$:

$$\boldsymbol{X}_{k+1} = \boldsymbol{X}_k + \boldsymbol{C}^{up}{}_k \left(\boldsymbol{C}^{pp}{}_k + \boldsymbol{\Gamma}\right)^{-1} \left(\boldsymbol{Z}_k - \hat{\boldsymbol{Y}}_k\right), \tag{A.7}$$

$$\boldsymbol{C}^{pp}{}_k = \frac{1}{J} \sum_{j=1}^{J} (\hat{y}_k^j - \overline{\hat{\boldsymbol{Y}}}_k)\,(\hat{y}_k^j - \overline{\hat{\boldsymbol{Y}}}_k)^{\mathsf{T}}, \tag{A.8}$$

$$\boldsymbol{C}^{up}{}_k = \frac{1}{J} \sum_{j=1}^{J} (\hat{\boldsymbol{x}}_k^j - \overline{\hat{\boldsymbol{X}}}_k)\,(\underbrace{\hat{\boldsymbol{Y}}_k^j}_{\boldsymbol{H}_k\,\hat{\boldsymbol{x}}_k^j} - \overline{\hat{\boldsymbol{Y}}}_k)^{\mathsf{T}}. \tag{A.9}$$

In this notation, the variables have unique values at a time step k, which is why the difference between $k|k-1$ and $k|k$ was dropped. Often the measurements $\boldsymbol{z}_k$ are not augmented: $\boldsymbol{Z}_k = [\boldsymbol{z}_k, \ldots, \boldsymbol{z}_k]$ to facilitate the analysis of the stability and well-posedness of the EnKF. However, perturbing the measurements $\boldsymbol{z}_k$, creating artificially individual observations $\boldsymbol{z}_k^j$ for every member $\boldsymbol{x}_k^j$, prevents particles from synchronizing and collapsing completely to a potentially suboptimal solution [KELL14]. The analysis of Eq. (A.7) - Eq. (A.9) has been performed in the continuous-time limit [SCHI18], i.e. the time between the discrete samples k is small enough with regard to the function behavior as if the measurements were continuously sampled. In this limit, they derive a system of ordinary differential equations (ODEs) for the analysis.

A.2 Algorithms

A.2.1 Recursive least squares (RLS)

The RLS algorithm is an iterative least squares fit making use of the estimate of the previous time instance. In the case of identifying a mechanistic force model, it is a recursive formulation of the method of instantaneous undeformed chip thickness.

Assuming an arbitrary measurement system with a model output y_k that has a nonlinear relationship to the estimated state vector $\hat{\boldsymbol{x}}_k$ for time instance k [STRE79]. For a proper state space representation, the relationship is linearized through the measurement matrix $\boldsymbol{M}_k^\mathsf{T}$:

$$y_k = f(\boldsymbol{x}_k) \approx \boldsymbol{M}_k^\mathsf{T}\, \boldsymbol{x}_k. \tag{A.10}$$

The model-matrix $\boldsymbol{M}_k^\mathsf{T}$ weights the error between the measurement and the prediction through the gain $\boldsymbol{G}_k$:

$$\hat{\boldsymbol{x}}_k = \hat{\boldsymbol{x}}_{k-1} + \boldsymbol{G}_k\, \left(\boldsymbol{z}_k - f(\hat{\boldsymbol{x}}_k)\right), \tag{A.11}$$

$$\boldsymbol{G}_k = \frac{\boldsymbol{P}_{k-1}\, \boldsymbol{M}_k}{\kappa + \boldsymbol{M}_k^\mathsf{T}\, \boldsymbol{P}_{k-1}\, \boldsymbol{M}_k}, \tag{A.12}$$

$$\boldsymbol{P}_k = \frac{1}{\kappa}\, \left[\boldsymbol{I} - \boldsymbol{G}_k\, \boldsymbol{M}_k^\mathsf{T}\right]\, \boldsymbol{P}_{k-1}. \tag{A.13}$$

The recursive approximation of the covariance matrix $\boldsymbol{P}_k$ limits the computational complexity. In the case on hand the state vector becomes $\boldsymbol{x}_k = [k_i \quad m_i]^\mathsf{T}$. The forgetting factor $\kappa \in (0, 1)$ weight the samples [STRE79] and was set to $\kappa = 0.98$. The initial value of the covariance matrix was set to $\boldsymbol{P}_0 = 10^5\, \boldsymbol{I}$ with the identity matrix $\boldsymbol{I}$.

A.2.2 Bounded optimization by quadratic approximation (BOBYQA)

BOBYQA is a local, constrained, derivative-free optimization algorithm. It is a constrained enhancement of the previously developed NEW UNCONSTRAINED OPTIMIZATION ALGORITHM (NEWUOA), both by POWELL [POWE04; POWE07; POWE09]. The bottleneck of both algorithms is the missing derivative. They approximate the nonlinear cost function with a quadratic function, resulting in a simple derivative. The idea is that a quadratic problem can be solved much faster than an arbitrary nonlinear problem. The algorithms use a local quadratic polynomial approximation Q of the nonlinear objective function J:

$$\begin{aligned} & Q \approx J, \\ \Rightarrow & Q(x_i) = J(x_i), \qquad \text{where} \qquad i \in \{1, 2, \ldots, m\}. \end{aligned} \tag{A.14}$$

The two functions coincide at $m = 2\,n + 1$ points, where the number m of supporting points for the approximation depends on the number of optimization variables n. The now quadratic objective function Q is minimized within a trust region, which represents the reliability of the function Q with regard to the original function J. To construct this approximation, it maintains the set of m interpolation points throughout the iterations. The points – and with them the function Q – are updated sequentially over the iterations of the algorithm. Since the supporting points do not fully define the approximated model Q, the remaining parameters are limited by minimizing the FROBENIUS norm of the change in the second derivative of the model.

BOBYQA includes an advanced method to overcome numeric rounding errors, which can occasionally cause damage to the optimization leading to failure [POWE09]. It solves a subproblem in each iteration, where the trust region radius d_k is restricted to the external box constraints:

$$\boldsymbol{x}_{lb} \leq \boldsymbol{x}_k + d_k \leq \boldsymbol{x}_{ub}. \tag{A.15}$$

A study on different optimization algorithms for identifying this process model has shown that BOBYQA is among the fastest and the most accurate choice for this case [SCHW19a]. In the case of the KIENZLE-model, the constant coefficients k_t and k_r have almost linear influence (it is cubic but at a very low rate). The exponential coefficients m_t and m_r exhibit a nonlinear influence. The nonlinearity is even higher in the KLINE & DEVOR-model.

A.2.3 KALMAN filter (KF)

"It is much more than an estimator, because it propagates the current state of knowledge of the dynamic system, including the statistical influence of random dynamic perturbations and the effects of all past measurements" [GREW15, p. 3].

The KALMAN filter – in continuous time also referred to as KALMAN-BUCY filter [SCHR77, p. 44] – is an optimal linear filter. It is statistically optimal in terms of a quadratic function of the estimation error [GREW15, p. 1]. Developed in

1960 by RUDOLF E. KALMAN [KALM60], the KF estimates a linear, discrete, time-invariant system in the presence of GAUSSIAN noise. If the noise is non-GAUSSIAN, the KF may not obtain the optimal solution but it provides still the best linear solution [SIMO06, p. 130]. The KF has a particular strength in fusing different information sources. It combines sensor signals of different GAUSSIAN uncertainties to a more accurate estimation.
The filter uses the state space formulation to estimate linear dependent states considering all past observation [MAYB79]. Assuming a linear, discrete-time system with uncorrelated white GAUSSIAN noise $\boldsymbol{w}_{k-1}$, $\boldsymbol{v}_k$, the KF minimizes the variance of the expected measurements:

$$\begin{aligned} \boldsymbol{x}_k &= \boldsymbol{A}_{k-1}\,\boldsymbol{x}_{k-1} + \boldsymbol{B}_{k-1}\,u_k + \boldsymbol{w}_{k-1}, \\ y_k &= \boldsymbol{H}_k\,\boldsymbol{x}_k + \boldsymbol{v}_k. \end{aligned} \tag{A.16}$$

Demanding white noise implies that it is not correlated in time and has equal power (amplitude) [MAYB79]. White noise is so pervasive in real-world electronic systems that it is almost surely in any sensor signal [GREW15, p. 148]. A KF solves a linear quadratic GAUSSIAN estimation problem: the dynamic system is linear, the cost function is quadratic, the random process is GAUSSIAN [GREW15, p. 132].

The filter works in two steps, a prediction step (Eq. (A.17)) and an update step (Eq. (A.20)):

$$\hat{\boldsymbol{x}}_{k|k-1} = \boldsymbol{A}_{k-1}\,\boldsymbol{x}_{k-1|k} + \boldsymbol{B}_{k-1}\,u_k, \tag{A.17}$$

$$\boldsymbol{P}_{k|k-1} = \boldsymbol{A}_{k-1}\,\boldsymbol{P}_{k-1|k-1}\,\boldsymbol{A}_{k-1}^{\mathsf{T}} + \boldsymbol{Q}_{k-1}. \tag{A.18}$$

First, the state $\hat{\boldsymbol{x}}_{k|k-1}$ is estimated for time step k based on the information of the previous time step $k-1$. The covariance $\boldsymbol{P}_k$ of the estimated state vector $\hat{\boldsymbol{x}}_k$ describes the linear relation between the variables of the state vector. The system noise covariance matrix $\boldsymbol{Q}_k$ specifies how much to trust the model $\boldsymbol{A}_k$. If the system noise covariance is close to zero $\boldsymbol{Q}_k = \mathbf{0}$, the model is considered to be ideal.

The update step first determines the KALMAN gain $\boldsymbol{G}_k$ from the state covariance $\boldsymbol{P}_k$ and the measurement covariance $\boldsymbol{R}_k$. The higher $\boldsymbol{R}_k$, the more noise is expected to be in the measurements $\boldsymbol{z}_k$. Note that the KF requires the measurement to be corrupted by at least some noise $\boldsymbol{R}_k > \mathbf{0}$ [SIMO06, p. 83]:

$$G_k = P_{k|k-1}\, H_k^\mathsf{T} \left[H_k P_{k|k-1}\, H_k^\mathsf{T} + R_k \right]^{-1}, \tag{A.19}$$

$$\hat{x}_{k|k} = \hat{x}_{k|k-1} + G_k \left[z_k - H_k \hat{x}_{k|k-1} \right], \tag{A.20}$$

$$P_{k|k} = \left[I - G_k\, H_k \right] P_{k|k-1}. \tag{A.21}$$

The predicted state estimate $\hat{x}_{k|k-1}$ is updated based on the measurements z_k weighted by the KALMAN gain G_k. The difference between the expected (measurable) states $H_k\hat{x}_{k|k-1}$ and the actual measurements z_k is called innovation [GREW15, p. 424]. At the end, the state covariance matrix P_k is updated. If the state vector x_k is constant, the KF reduces to a recursive least-squares optimization [SIMO06, p. 129].

Tuning of the KF consists of defining the covariance matrices Q_k and R_k. If the measurement noise is uncorrelated, R_k is a diagonal matrix [GREW15, p. 141]. The same is true for the system noise covariance Q_k. If the dynamic system does not change with time, i.e. it is time-invariant, the system matrices A, B, H, Q, and R are constants that is indicated by omitting the subscript $_k$.

A.3 Supplement

A.3.1 Complementary figures to Chapter 4

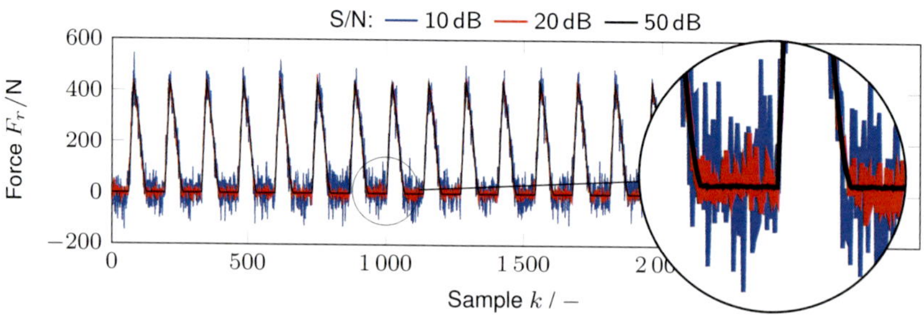

Figure A.1: *Simulated radial force signals F_r with added white,* GAUSSIAN *noise.*

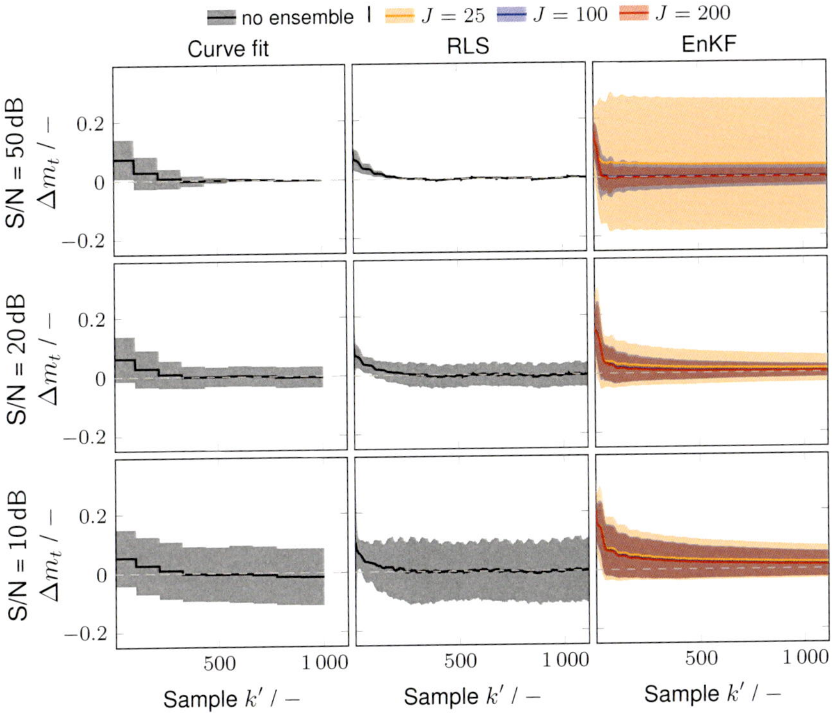

Figure A.2: *Evolution of the average error (—) of the exponential coefficient Δm_t and its spread ($\pm 2\,\sigma$, shaded areas ▒) of 500 simulations with random initial ensembles but identical ideal model coefficients (supplementary figure for the radial component: Fig. A.5)*

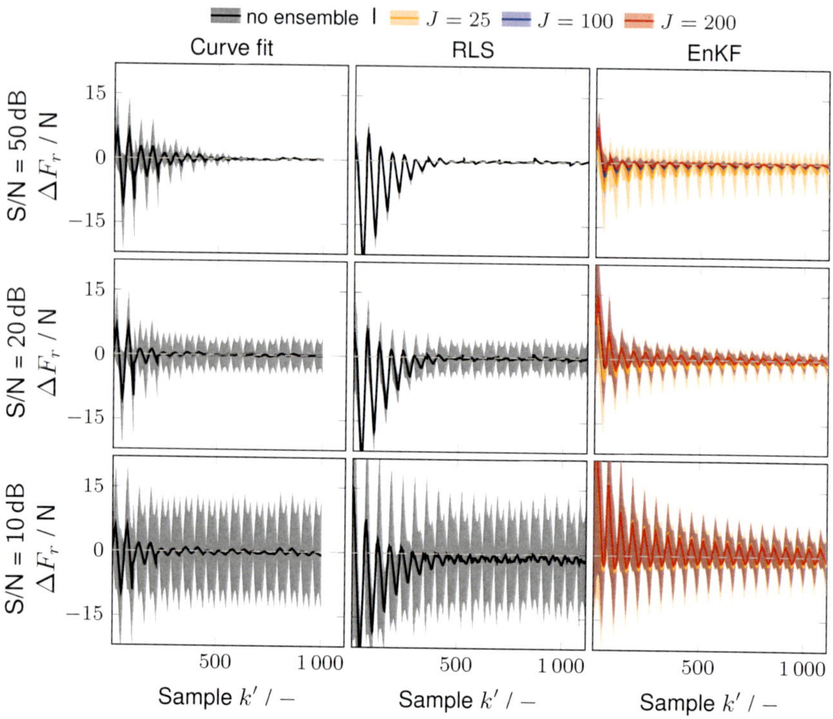

Figure A.3: *Evolution of the average error (—) of the radial component of the force ΔF_r and its spread ($\pm 2\,\sigma$, shaded areas) of 500 simulations with random initial ensembles but identical ideal model coefficients*

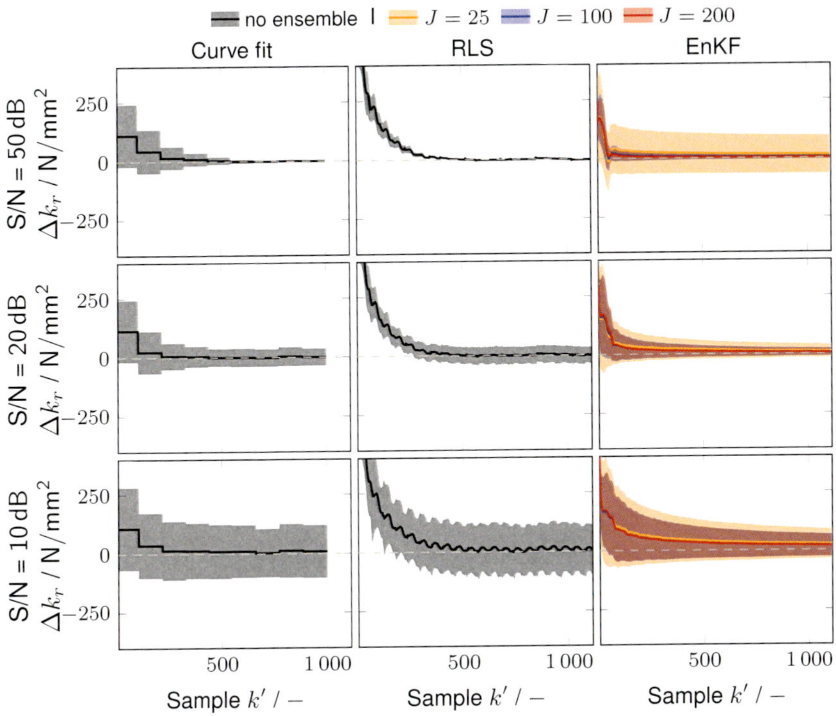

Figure A.4: *Evolution of the average error (—) of the linear coefficient Δk_r and its spread ($\pm 2\,\sigma$, shaded areas ▇) of 500 simulations with random initial ensembles but identical ideal model coefficients*

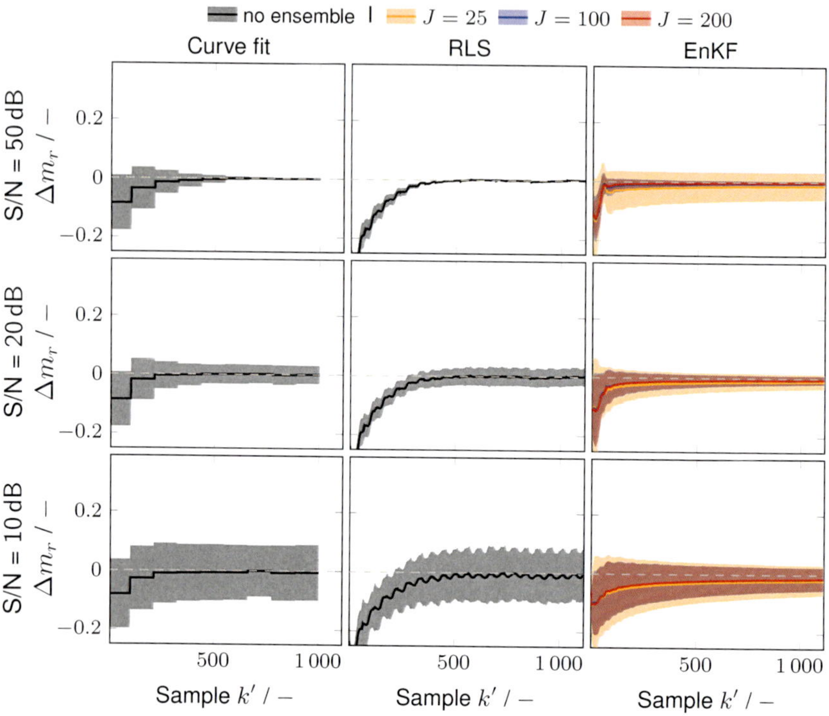

Figure A.5: *Evolution of the average error (—) of the exponential coefficient Δm_r and its spread ($\pm 2\,\sigma$, shaded areas) of 500 simulations with random initial ensembles but identical ideal model coefficients*

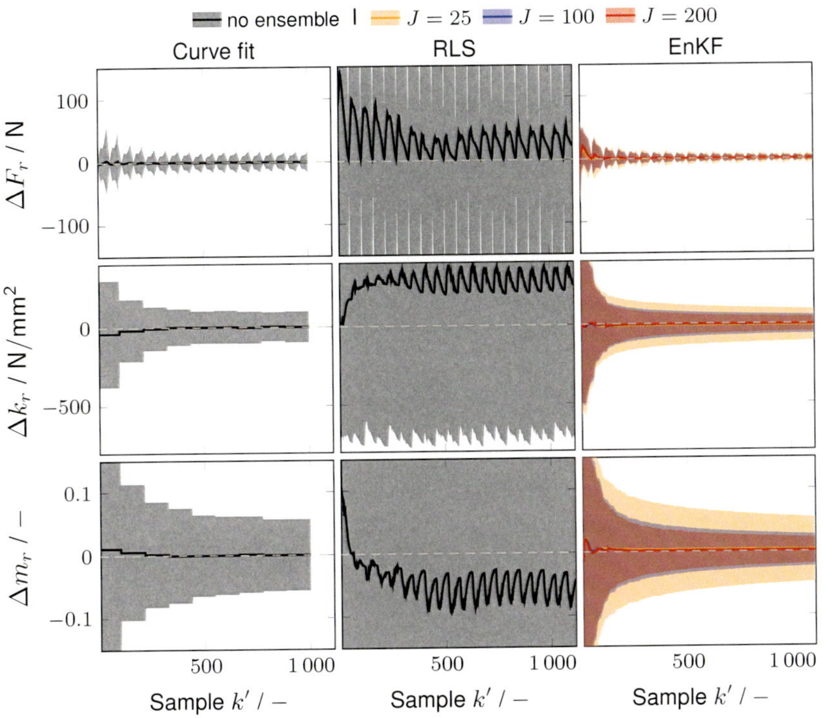

Figure A.6: *Evolution of the error (average —, $\pm 2\,\sigma$ ▬) of the force (top), the linear (middle), and the exponential coefficient (bottom) of the* KIENZLE*-model for 500 random simulations (artificial noise S/N = 20 dB)*

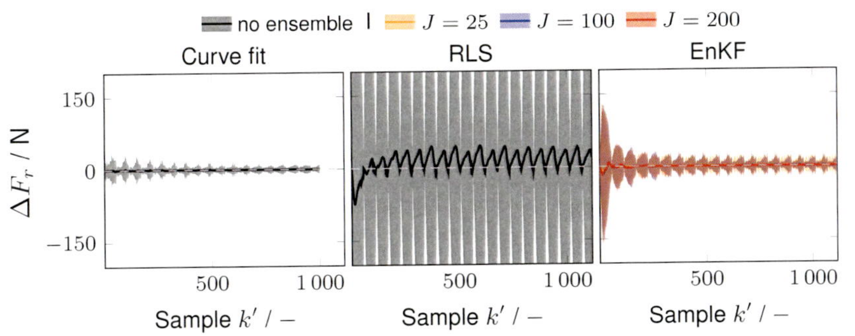

Figure A.7: *Evolution of the error (average —, $\pm 2\,\sigma$ ▬) in the force when identifying the* KLINE *&* DEVOR*-model in polar formulation (500 random simulations, artificial noise S/N = 20 dB)*

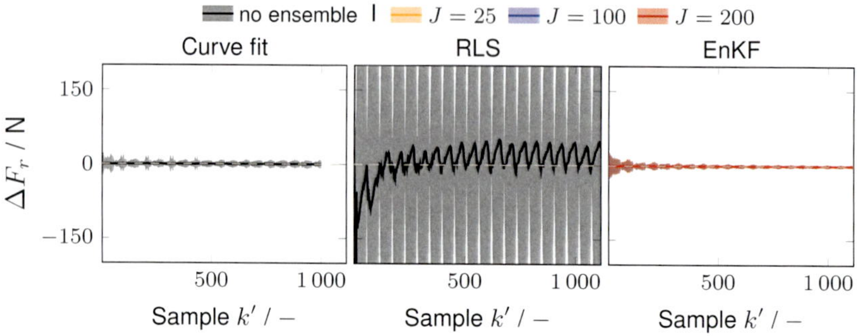

Figure A.8: *Evolution of the error (average —, $\pm 2\,\sigma$ ▬) in the force when identifying the* KLINE & DEVOR*-model in* CARTESIAN *formulation (500 random simulations, artificial noise S/N = 20 dB)*

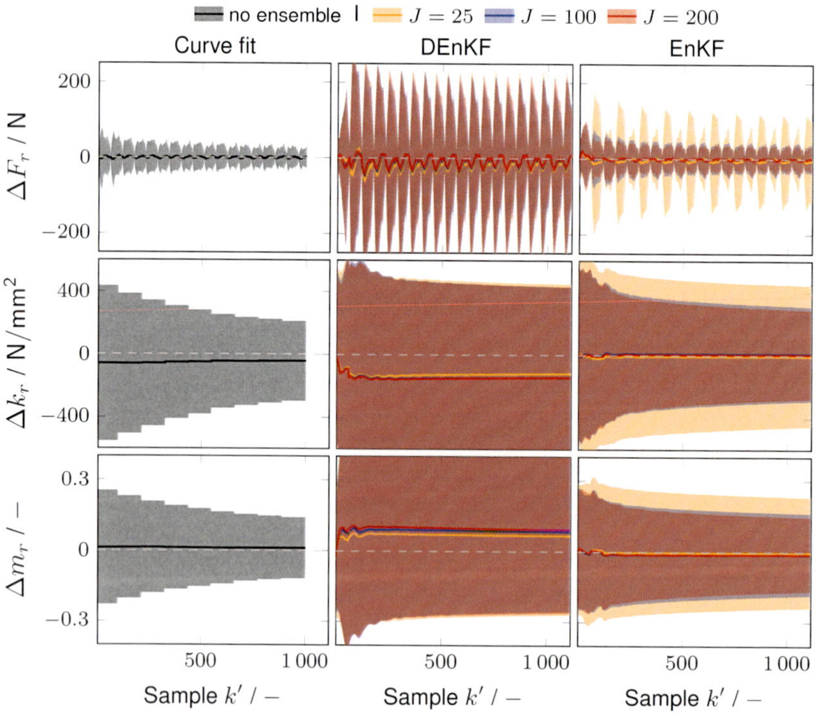

Figure A.9: *Evolution of the error (average —, $\pm 2\,\sigma$ ▬) of the radial components of the* KIENZLE*-model for 500 random simulations of a simultaneous identification of the* KIENZLE*-model and the* KLINE & DEVOR*-model in its polar formulation (S/N = 20 dB)*

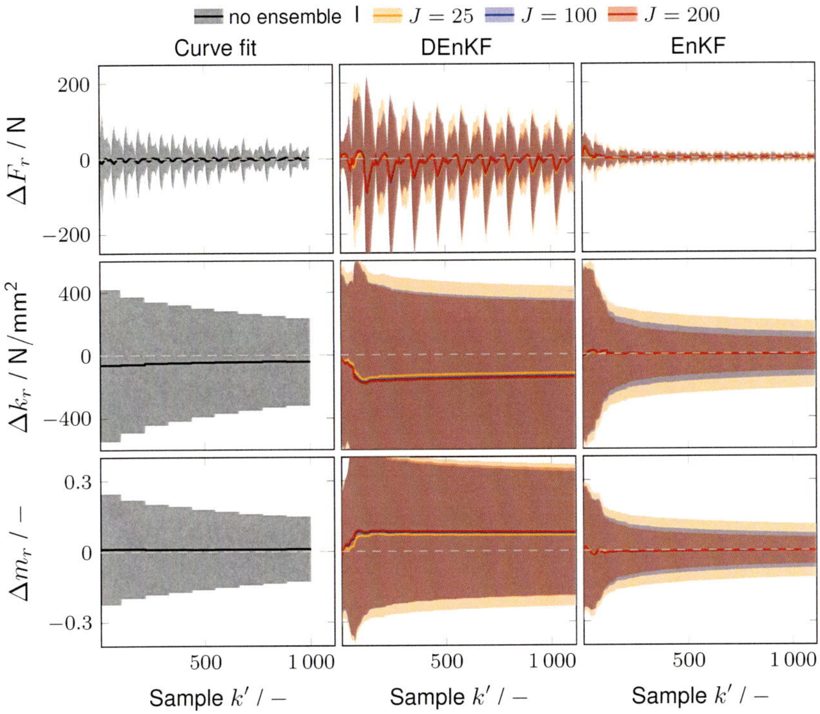

Figure A.10: *Evolution of the error (average —, $\pm 2\,\sigma$) in the radial components of the* KIENZLE*-model for 500 random simulations of a simultaneous identification of the* KIENZLE*-model and the* KLINE & DEVOR*-model in* CARTESIAN *formulation (S/N* = 20 dB*)*

A.3.2 Complementary figures to Chapter 5

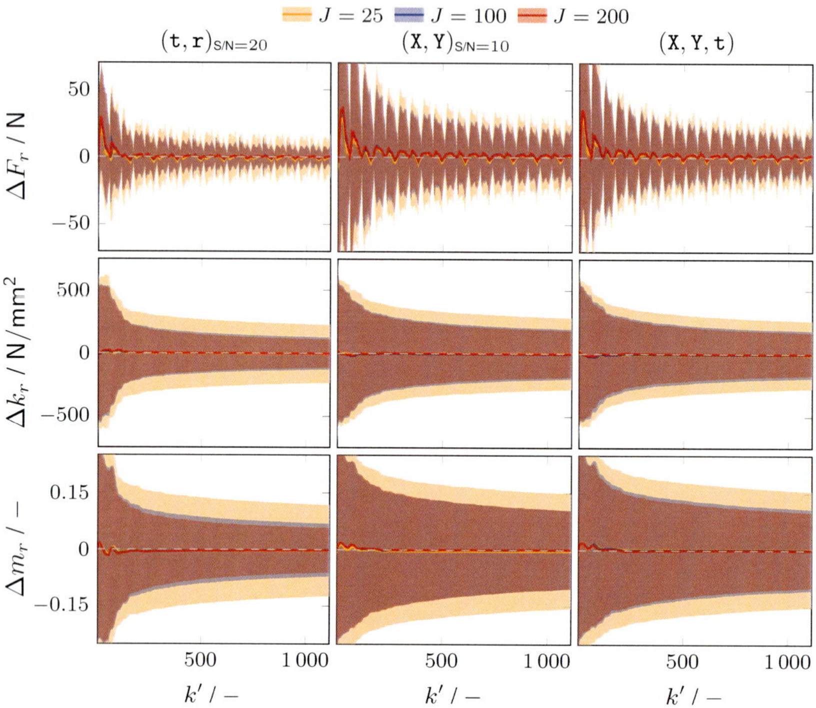

Figure A.11: *Evolution of the error of the radial force component F_r and coefficients of the* KIENZLE*-model; completing Fig. 5.2.*

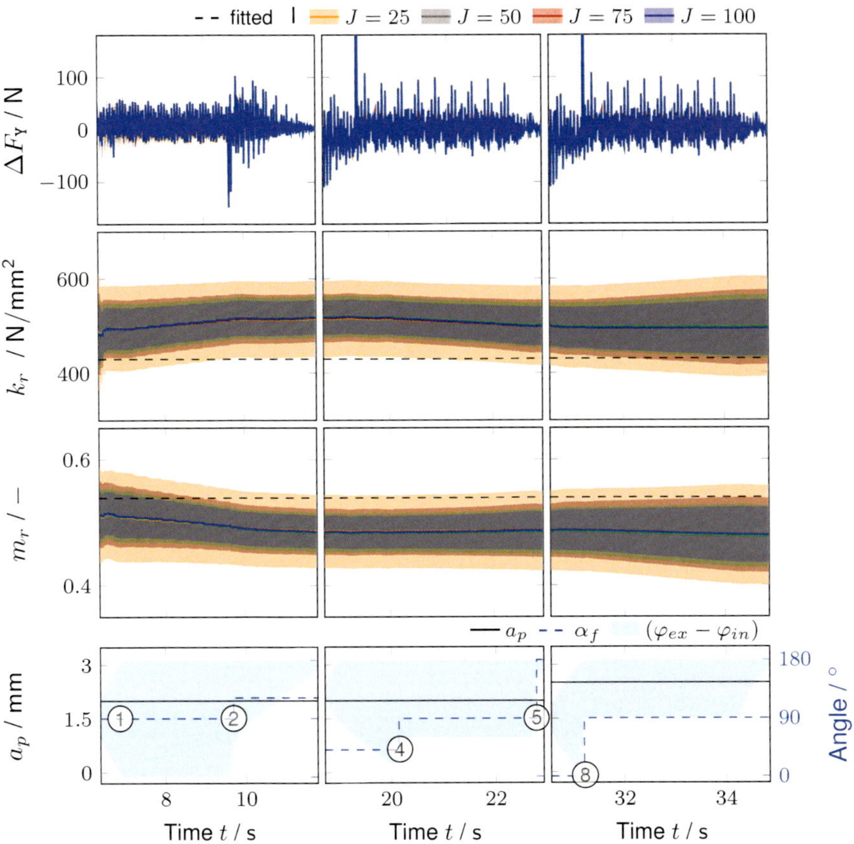

Figure A.12: *Evolution of 500 random simulations identifying the radial deviation model. The figure presents the radial components completing Fig. 5.6.*

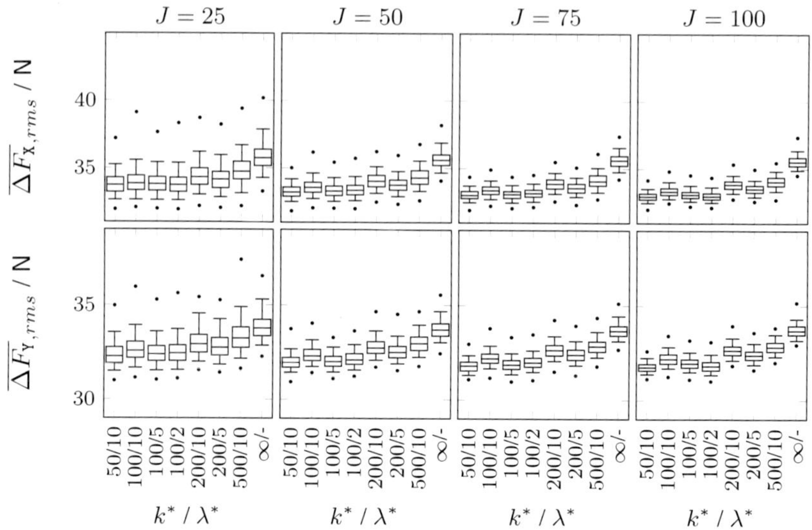

Figure A.13: *Boxplots of the summed-up rms error between identified force model and measurements for different combinations of inflation steps k^* and inflation fractions λ^*. The parameters of the* KLINE *&* DEVOR*-model were expressed in polar coordinates.*

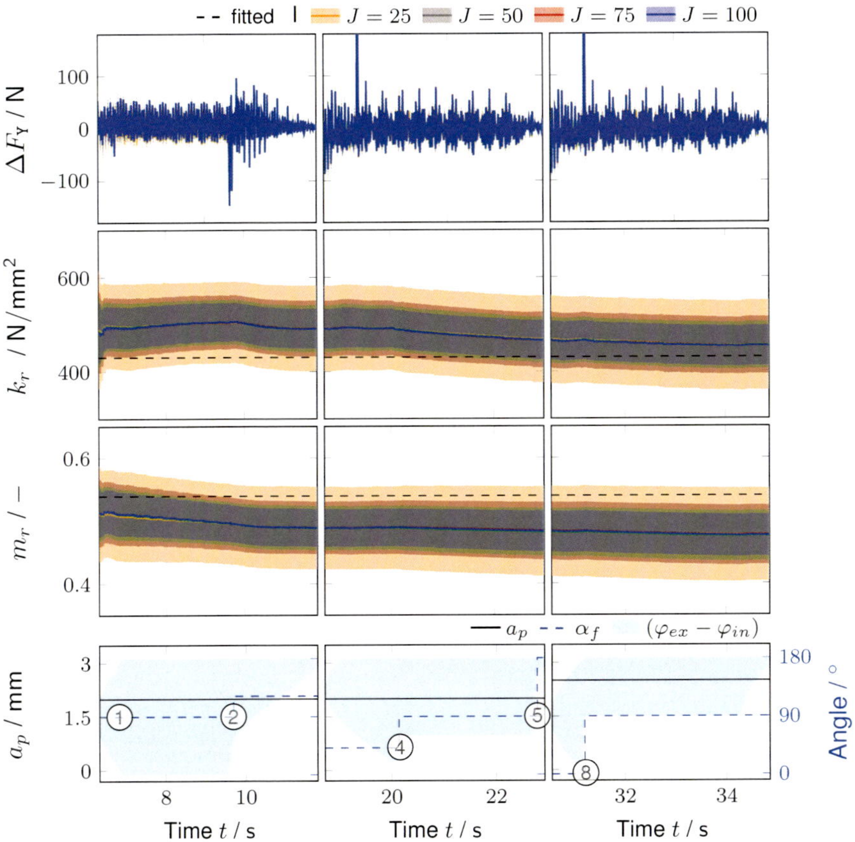

Figure A.14: *Evolution for 500 random simulations on real measurements. The figure completes Fig. 5.8.*

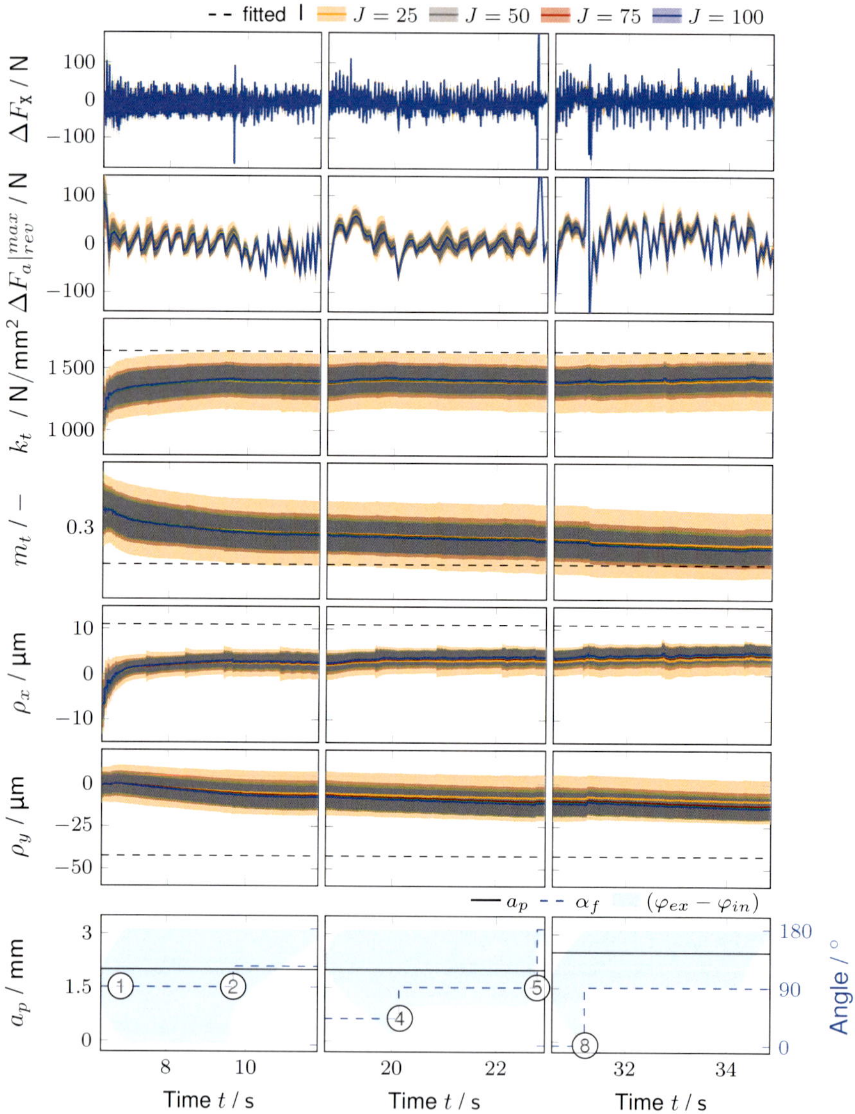

Figure A.15: *Evolution of the identification on real measurements for 500 random initial ensembles (supplementary figure of the radial force: Fig. A.16)*

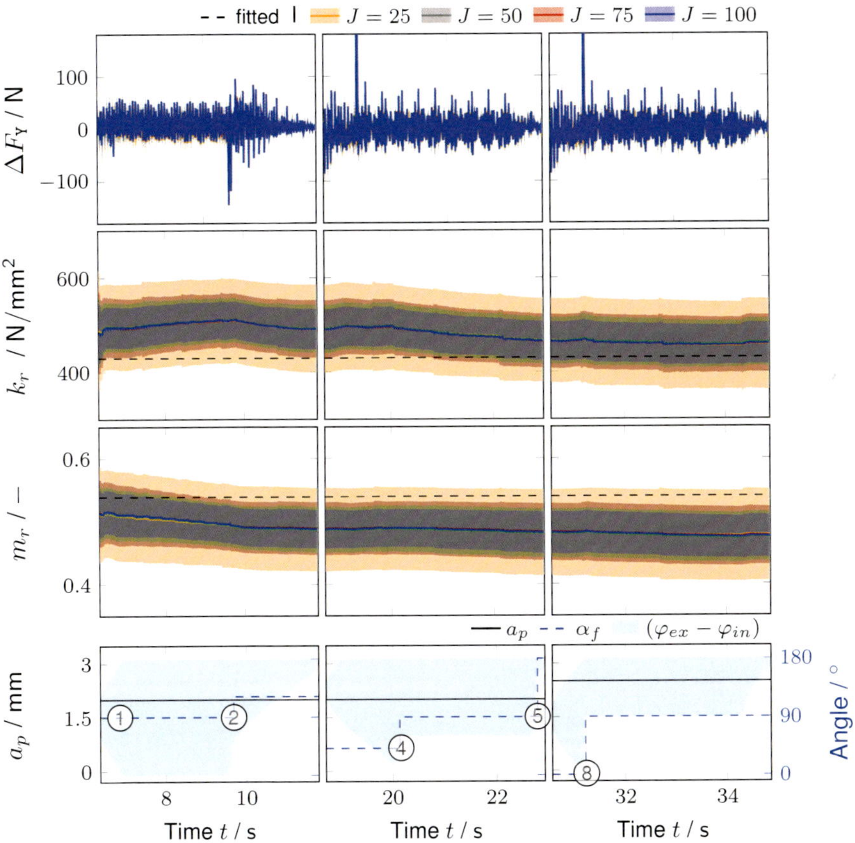

Figure A.16: *Evolution for 500 random simulations on real measurements. The figure completes Fig. A.15.*

A.3.3 Complementary figures to Chapter 6

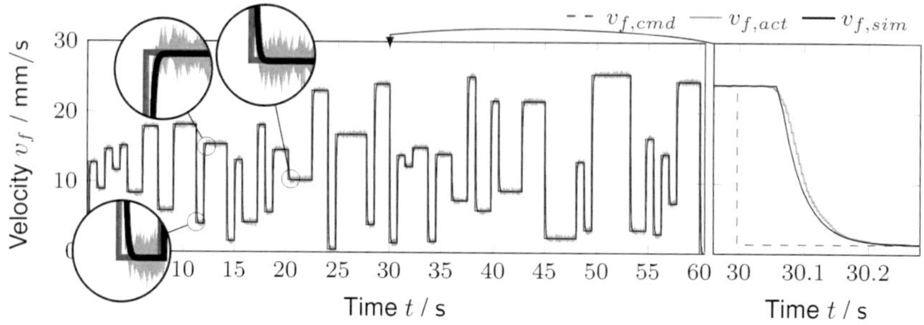

Figure A.17: *Reaction of the Y-axis of the* MAZAK VARIAXIS I600 *to step changes of the commanded velocity modeled as a second-order system (*$K = 0.9978$, $\zeta = 1.5552$, $\omega_0 = 80.5162$, $T_d = 60$ ms*)*

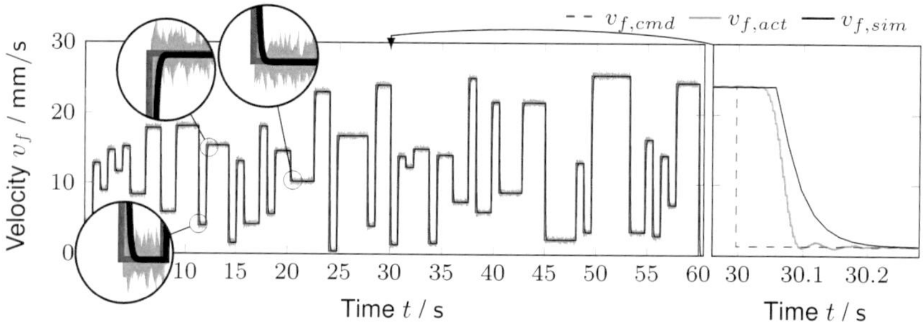

Figure A.18: *Reaction of the X- and Y-axis of the* MAZAK VARIAXIS I600 *with activated feedforward control to step changes of the commanded velocity modeled as a second-order system (*$K = 0.9978$, $\zeta = 1.5552$, $\omega_0 = 80.5162$, $T_d = 60$ ms*)*

A.3.4 Complementary figures to Chapter 7

(next page)

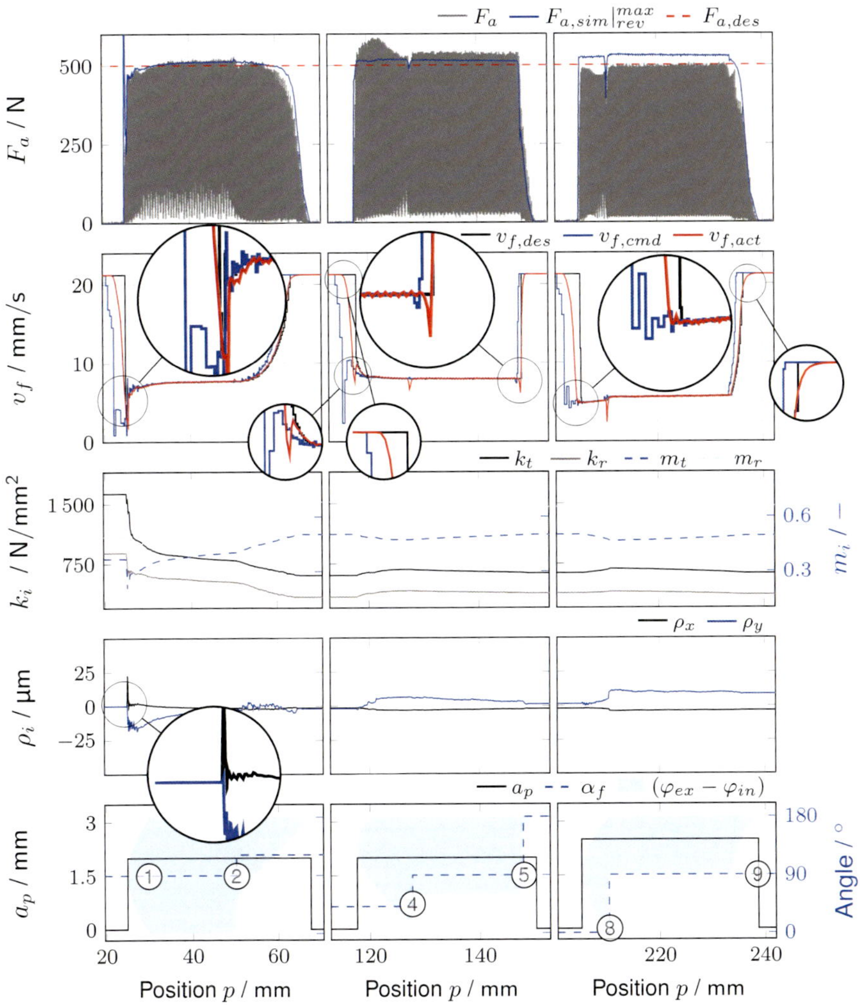

Figure A.19: *MPC with EnKF ($J = 50$, uniformly distributed initial ensemble $\boldsymbol{X}_0$, S/N=15 dB assumed to determine $\boldsymbol{\Gamma}$); displayed over the tool path (position domain)*

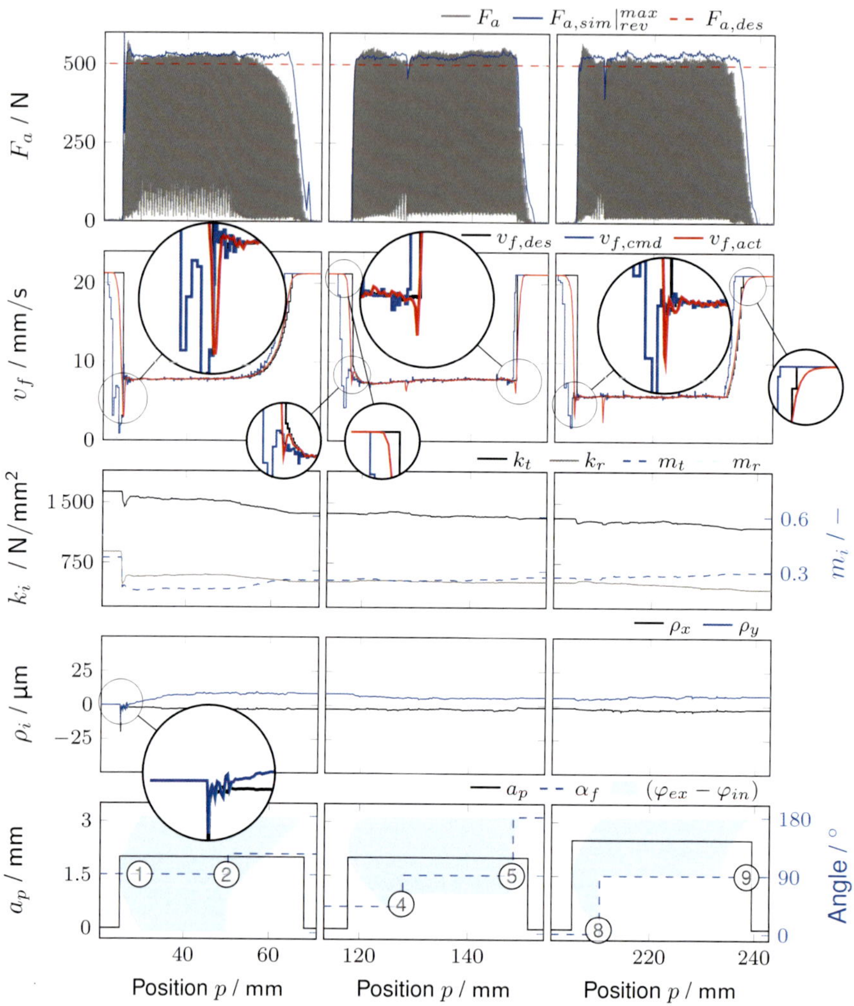

Figure A.20: *MPC with EnKF ($J = 50$, uniformly distributed $\mathbf{X}_0$, ensemble inflation); displayed over the tool path (position domain)*

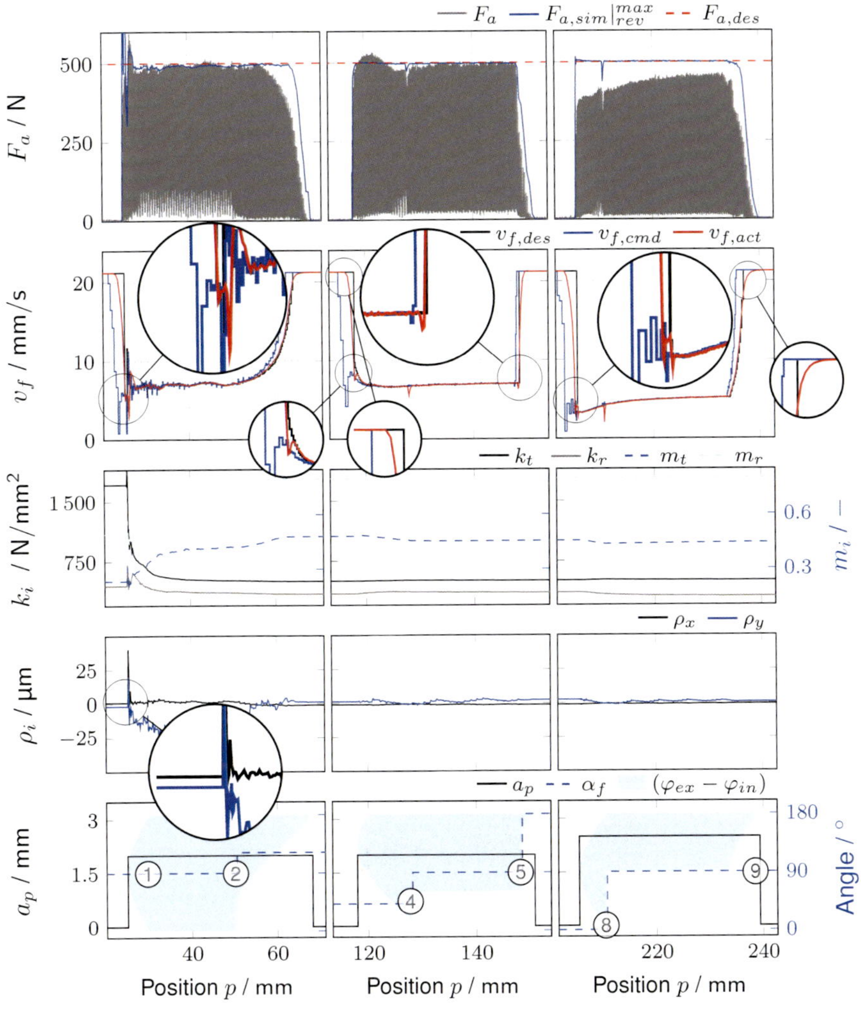

Figure A.21: *MPC with EnKF ($J = 50$,* GAUSSIAN *distributed initial ensemble around $k_t = 1700\,\text{N/mm}^2$, $m_t = 0.18$, $k_r = 350\,\text{N/mm}^2$, $m_r = 0.55$, S/N=15 dB assumed to determine $\mathbf{\Gamma}$); displayed over the tool path (position domain)*

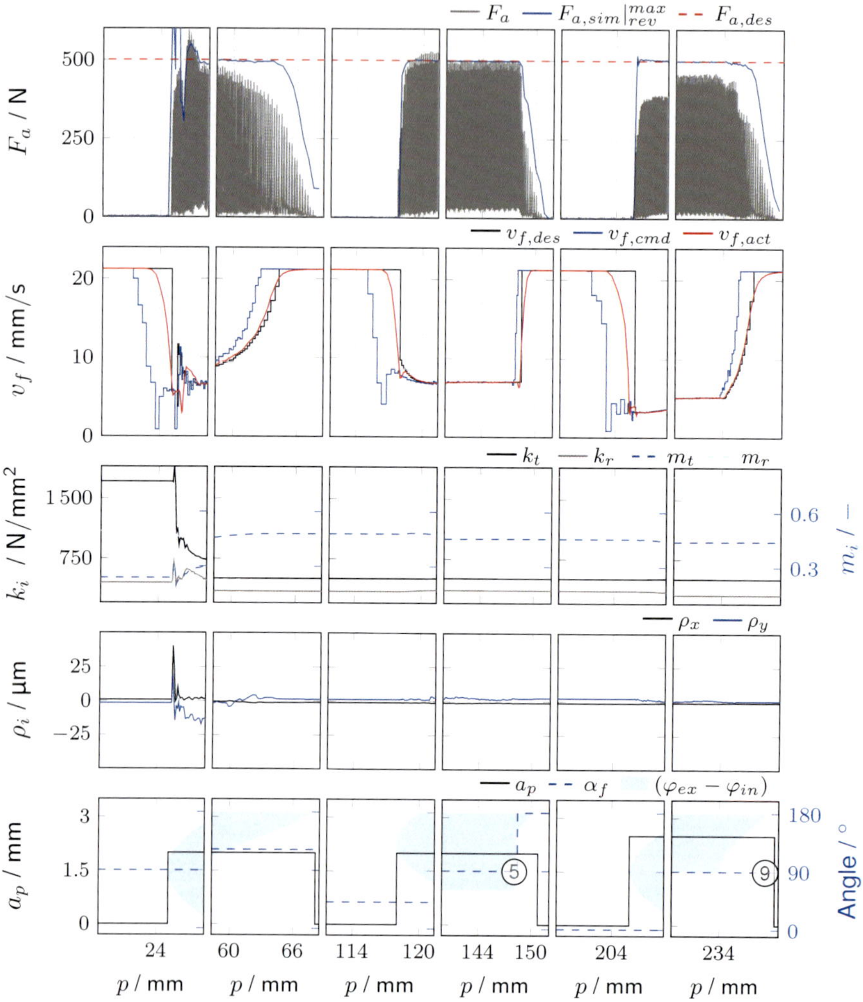

Figure A.22: *Regions of changing depth of cut a_p of Fig. 7.6 but in position domain (MPC, EnKF with* GAUSSIAN *$\boldsymbol{X}_0$)*

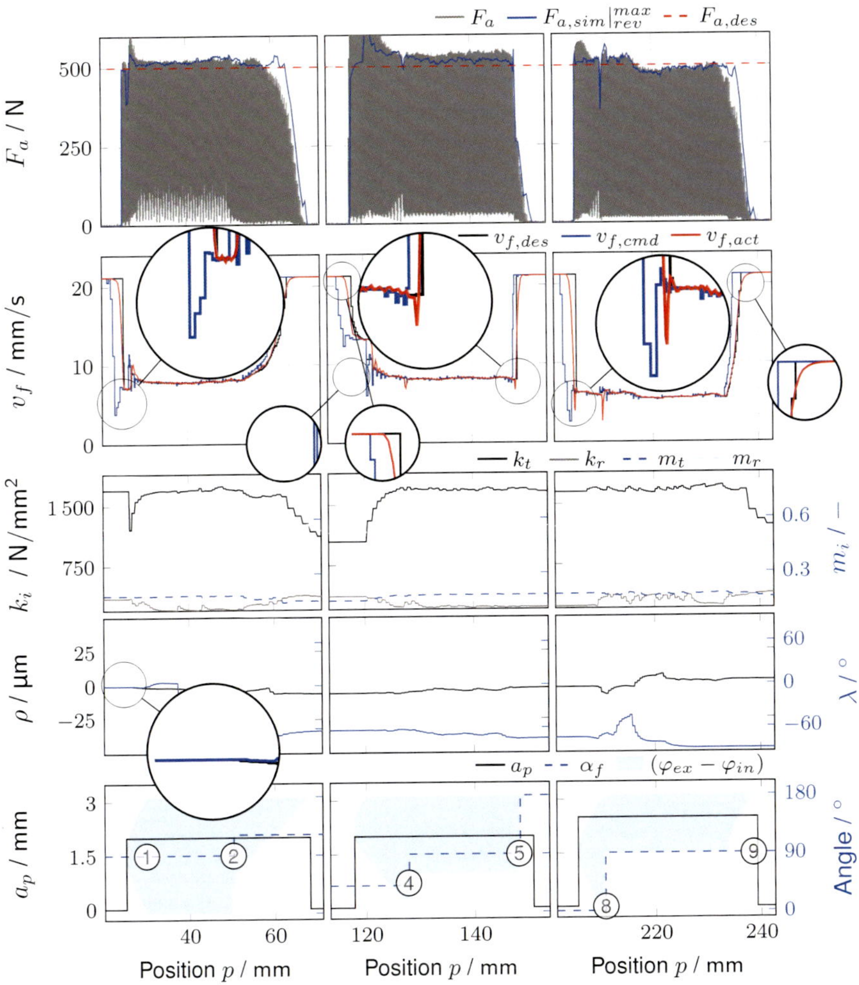

Figure A.23: *MPFC with a curve fit for model identification in accordance to [ADAM16a; SCHW19a; STEM20]*

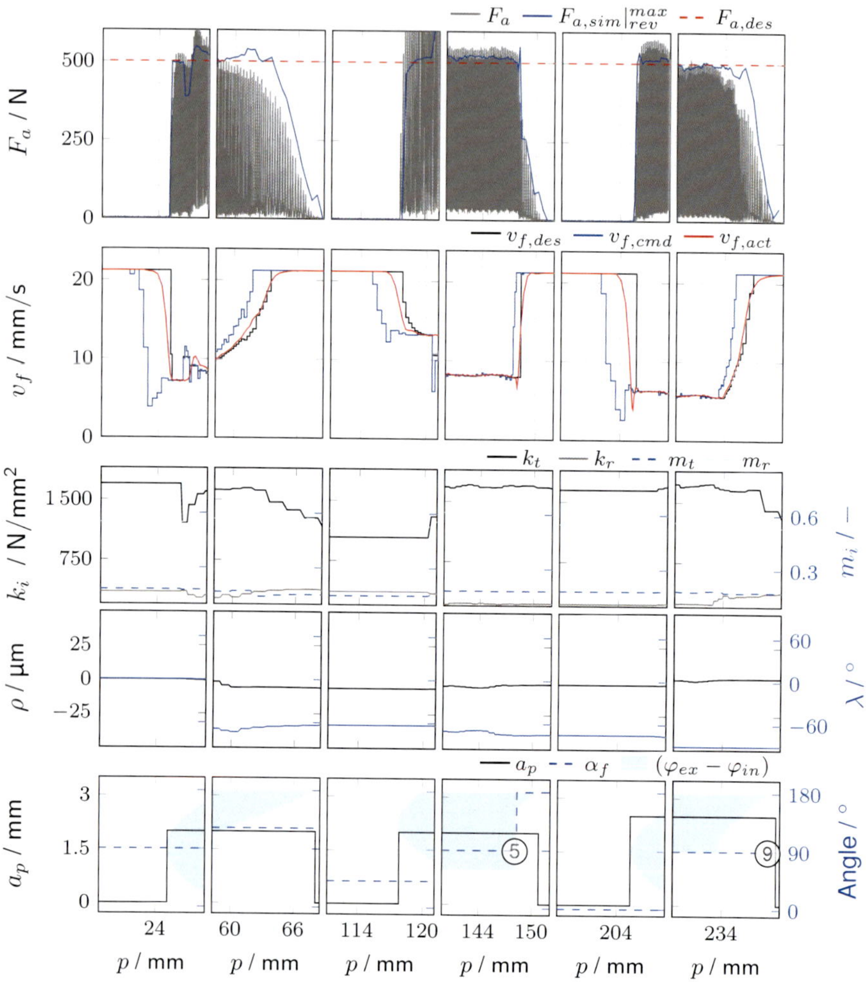

Figure A.24: *Regions of changing depth of cut a_p of Fig. 7.8 but in position domain (MPC, curve fit every* 80 ms*)*

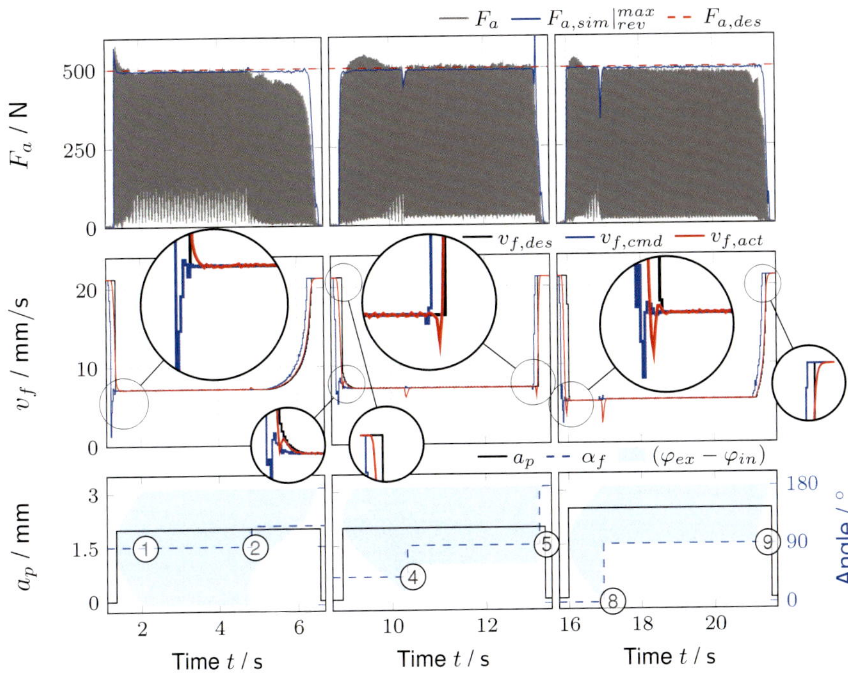

Figure A.25: *MPC for velocity control. Velocity reference determined based on [KÖNI82, p. 95] ($k_t = 1700\,\mathrm{N/mm^2}$, $m_t = 0.18$, $k_r = 350\,\mathrm{N/mm^2}$, $m_r = 0.55$).*

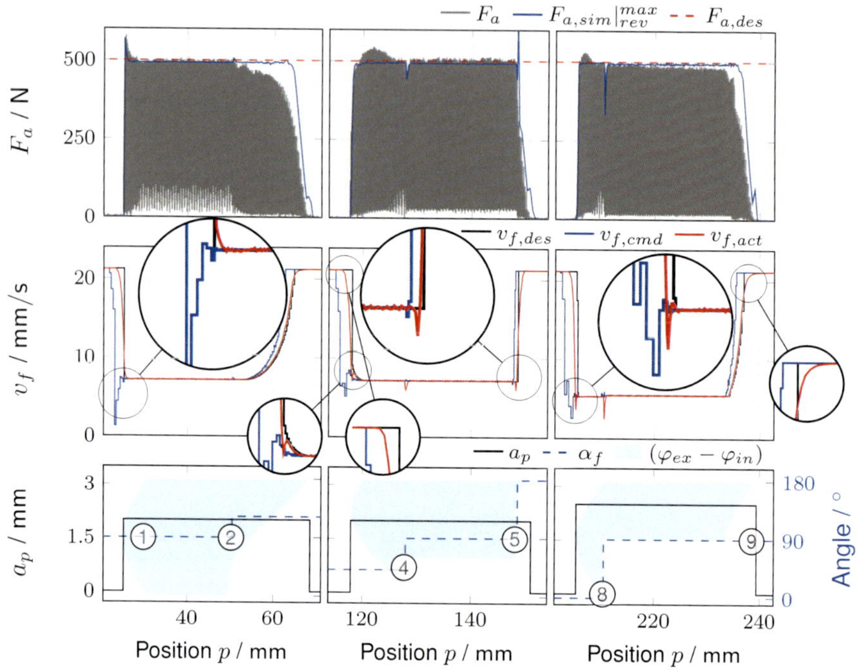

Figure A.26: *MPC for velocity control. Velocity reference determined based on [KÖNI82, p. 95] (k_t = 1700 N/mm^2, $m_t = 0.18$, k_r = 350 N/mm^2, $m_r = 0.55$, no radial deviation), displayed over the tool path (position domain).*

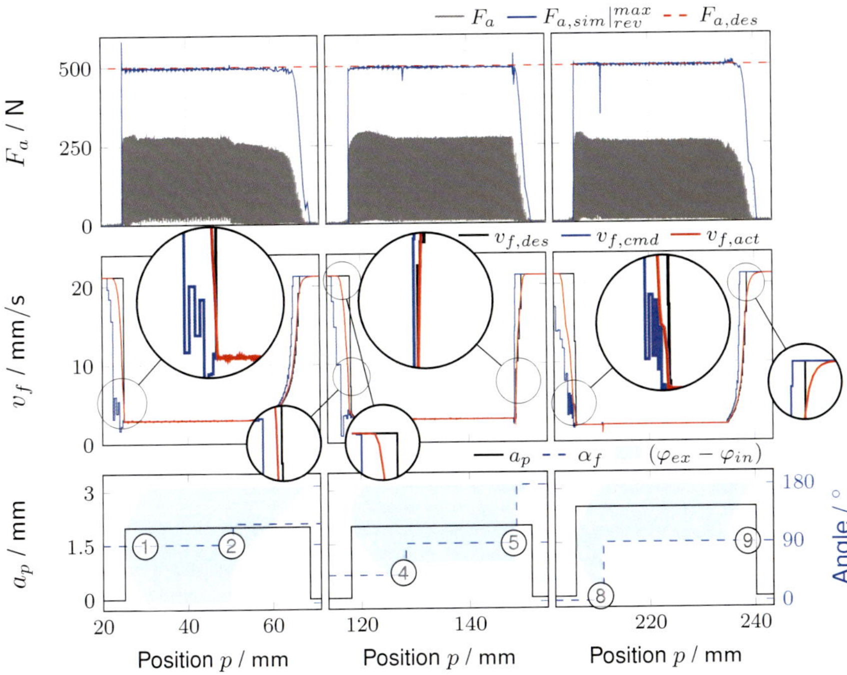

Figure A.27: *MPC for velocity control. Velocity reference determined based on the initial value of an uniformly distributed ensemble (k_t = 1762 N/mm^2, m_t = 0.236, k_r = 447 N/mm^2, m_r = 0.494, no radial deviation), displayed over the tool path (position domain).*

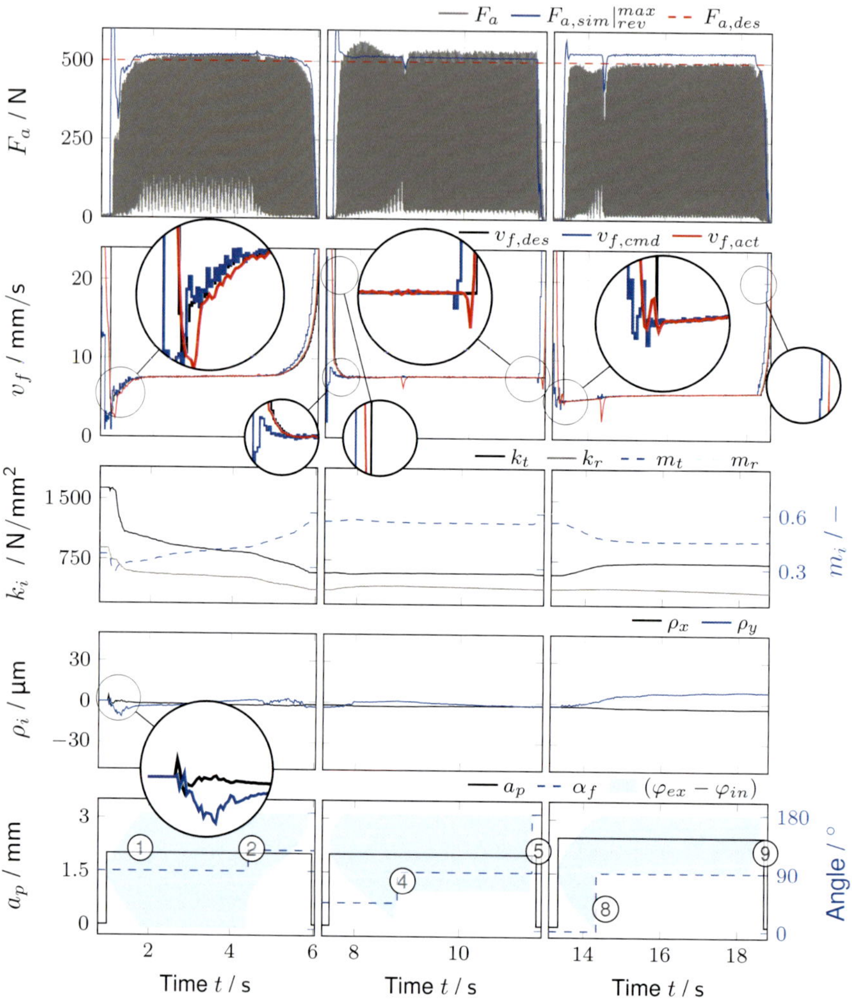

***Figure A.28:** MPC with maximum allowed feed per tooth of $f_{t,max} = 0.39$ mm and with EnKF for identification ($J = 50$, uniformly distributed initial ensemble, S/N=15 dB assumed to determine $\boldsymbol{\Gamma}$)*

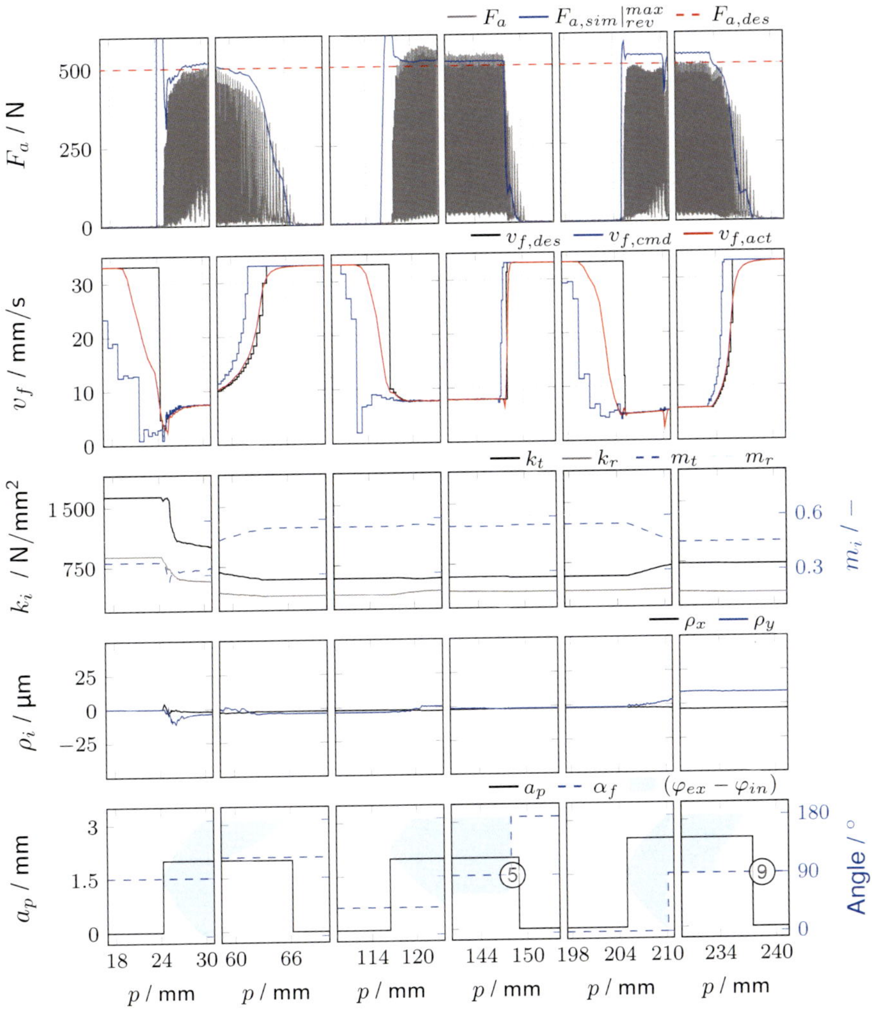

Figure A.29: *Regions of changing depth of cut a_p of Fig. A.28 but in position domain (MPC with $f_{t,max} = 0.39$ mm, EnKF with uniform $\boldsymbol{X}_0$)*

Bibliography

[ABEL06] D. Abel & A. Bollig: *Rapid control prototyping: Methoden und Anwendungen.* 1. edt. Berlin: Springer, 2006. 400 pp. ISBN: 978-3-540-29524-2.

[ADAM16a] O. Adams: *Modellbasierte prädiktive Kraftregelung beim Fräsen.* 1. edt. Ergebnisse aus der Produktionstechnik 2016, Band 29. Aachen: Apprimus Verlag, 2016. 143 pp. ISBN: 978-3-86359-455-8.

[ADAM16b] O. Adams; F. Klocke; M. Schwenzer; S. Stemmler Sebastian, & D. Abel: Model-based Predictive Force Control in Milling – System Identification. In: *3rd International Conference on System-Integrated Intelligence: New Challenges for Product and Production Engineering* 26 (2016), pp. 214–220. ISSN: 2212-0173. DOI: 10.1016/j.protcy.2016.08.029.

[ADEM15] K. A. M. Adem; R. Fales, & A. S. El-Gizawy: Identification of cutting force coefficients for the linear and nonlinear force models in end milling process using average forces and optimization technique methods. In: *The International Journal of Advanced Manufacturing Technology* 79.9 (Aug. 2015), pp. 1671–1687. ISSN: 0268-3768, 1433-3015. DOI: 10.1007/s00170-015-6935-3.

[AGGA13] S. Aggarwal; N. Nešić, & P. Xirouchakis: Cutting torque and tangential cutting force coefficient identification from spindle motor current. In: *The International Journal of Advanced Manufacturing Technology* 65.1 (Mar. 2013), pp. 81–95. ISSN: 0268-3768, 1433-3015. DOI: 10.1007/s00170-012-4152-x.

[ALTI17] Y. Altintas & D. Aslan: Integration of virtual and on-line machining process control and monitoring. In: *CIRP Annals - Manufacturing Technology* 66.1 (Jan. 1, 2017), pp. 349–352. ISSN: 0007-8506. DOI: 10.1016/j.cirp.2017.04.047.

[ALTI90] Y. Altintas & C. C. H. Ma: Direct Adaptive Control of Milling Force. In: *Proceedings of the IEEE International Workshop on Intelligent Motion Control* 1 (Aug. 20, 1990), pp. 13–20. DOI: 10.1109/IMC.1990.687288.

[ALTI94] Y. ALTINTAS: Direct adaptive control of end milling process. In: *International Journal of Machine Tools and Manufacture* 34.4 (May 1, 1994), pp. 461–472. ISSN: 0890-6955. DOI: 10.1016/0890-6955(94)90078-7.

[ALTI96] Y. ALTINTAS & P. LEE: A General Mechanics and Dynamics Model for Helical End Mills. In: *CIRP Annals - Manufacturing Technology* 45.1 (Jan. 1, 1996), pp. 59–64. ISSN: 0007-8506. DOI: 10.1016/S0007-8506(07)63017-0.

[AUER15] T. AUERBACH; S. GIERLINGS; D. VESELOVAC; R. SEIDNER; S. KAMPS, & F. KLOCKE: Concept for a Technology Assistance System to Analyze and Evaluate Materials and Tools for Milling. In: *Proceedings of the ASME Turbo Expo: Turbine Technical Conference and Exposition 2015*. ASME Turbo Expo. Vol. 6. ASME, June 15, 2015, V006T21A006. ISBN: 978-0-7918-5675-8. DOI: 10.1115/GT2015-43209.

[AUER18] T. AUERBACH: *Automatisierte, datenbasierte Wissensgenerierung beim Fräsen*. In collab. with R. AACHEN. 1. edt. Ergebnisse aus der Produktionstechnik Technologie der Fertigungsverfahren 2018, Band 28. Aachen: Apprimus Verlag, 2018. 168 pp. ISBN: 978-3-86359-646-0.

[AY18] M. AY; S. STEMMLER; D. ABEL; M. SCHWENZER, & F. KLOCKE: System Identification of a CNC Machining Center with Support Vector Machines. In: 2018 26th Mediterranean Conference on Control and Automation (MED). Zadar, Croatia: IEEE, June 2018, pp. 1–9. ISBN: 978-1-5386-7890-9. DOI: 10.1109/MED.2018.8442437.

[AY19a] M. AY; S. STEMMLER; M. SCHWENZER; D. ABEL, & T. BERGS: Model Predictive Control in Milling based on Support Vector Machines. In: *IFAC-PapersOnLine* 52.13 (2019), pp. 1797–1802. ISSN: 2405-8963. DOI: 10.1016/j.ifacol.2019.11.462.

[AY19b] M. AY; D. STENGER; M. SCHWENZER; D. ABEL, & T. BERGS: Kernel Selection for Support Vector Machines for System Identification of a CNC Machining Center. In: *IFAC-PapersOnLine* 52.29 (2019), pp. 192–198. ISSN: 2405-8963. DOI: 10.1016/j.ifacol.2019.12.643.

[BAUM19] J. BAUMANN; T. SIEBRECHT; P. WIEDERKEHR, & D. BIERMANN: The effect of runout errors on process forces and tool wear. In: *Procedia CIRP* 79 (2019), pp. 39–44. ISSN: 2212-8271. DOI: 10.1016/j.procir.2019.02.008.

[BERG10a] K. BERGEMANN & S. REICH: A localization technique for ensemble Kalman filters. In: *Quarterly Journal of the Royal Meteorological Society* (2010), pp. 700–707. ISSN: 0035-9009, 1477-870X. DOI: 10.1002/qj.591.

[BERG10b] K. BERGEMANN & S. REICH: A mollified ensemble Kalman filter. In: *Quarterly Journal of the Royal Meteorological Society* 136.651 (July 2010), pp. 1636–1643. ISSN: 0035-9009. DOI: 10.1002/qj.672.

[BERG18] B. BERGMANN & T. GROVE: Basic principles for the design of cutting edge roundings. EN. In: *CIRP Annals* 67.1 (2018), pp. 73–78. ISSN: 00078506. DOI: 10.1016/j.cirp.2018.04.019.

[BERN19] T. BERNERS: *Anforderungsspezifische Lageregelung von Vorschubantrieben in Werkzeugmaschinen.* 1st ed. Vol. 38/2019. Ergebnisse aus der Produktionstechnik. Aachen: Apprimus Verlag, 2019. 141 pp. ISBN: 978-3-86359-765-8.

[BLÖM18] D. BLÖMKER; C. SCHILLINGS; P. WACKER, & S. WEISSMANN: Well Posedness and Convergence Analysis of the Ensemble Kalman Inversion. In: *arXiv:1810.08463* (Oct. 19, 2018). DOI: 10.1088/1361-6420/ab149c.

[BRAN81] A. BRANDSTÄTTER: *Ein Adaptive-Control-Optimization-System für das Fräsen.* WBK-Forschungsberichte aus dem Institut für Werkzeugmaschinen und Betriebstechnik der Universität Karlsruhe 4. Berlin: Springer, 1981. 139 pp. ISBN: 978-3-540-10948-8.

[CABR16] G. F. CABRAL: *Modeling and Simulation of Tool Engagement and Prediction of Process Forces in Milling.* Aachen: Apprimus Verlag, 2016. 137 pp. ISBN: 978-3-86359-427-5.

[CAMP12] G. CAMPATELLI & A. SCIPPA: Prediction of Milling Cutting Force Coefficients for Aluminum 6082-T4. In: *Procedia CIRP* 1 (2012), pp. 563–568. ISSN: 2212-8271. DOI: 10.1016/j.procir.2012.04.100.

[CARR97] F. CARRILLO & F. ROTELLA: A robust adaptive controller for constant turning force regulation. In: *Control Engineering Practice* 5.12 (Dec. 1, 1997), pp. 1673–1682. ISSN: 0967-0661. DOI: 10.1016/S0967-0661(97)10021-1.

[CEN16] L. CEN; S. N. MELKOTE; J. CASTLE, & H. APPELMAN: A Wireless Force-Sensing and Model-Based Approach for Enhancement of Machining Accuracy in Robotic Milling. In: *IEEE/ASME Transactions on Mechatronics* 21.5 (Oct. 2016), pp. 2227–2235. ISSN: 1083-4435, 1941-014X. DOI: 10.1109/TMECH.2016.2567319.

[CERA21] CERATIZIT S.A.: *Full process control with ToolScope*. 2021. URL: https://cuttingtools.ceratizit.com/gb/en/services/toolscope.html (visited on 01/03/2021).

[CGTE21] CGTECH INC.: *OptiPath*. 2021. URL: https://www.cgtech.com/products/product-line/100-optipath.html (visited on 01/03/2021).

[CHAD19] N. K. CHADA; C. SCHILLINGS, & S. WEISSMANN: On the Incorporation of Box-Constraints for Ensemble Kalman Inversion. In: *arXiv:1908.00696* (Oct. 14, 2019).

[CHAE07] J. CHAE & S. PARK: High frequency bandwidth measurements of micro cutting forces. In: *Selected papers from the 2nd International Conference on High Performance Cutting2nd CIRP International Conference on High Performance Cutting* 47.9 (July 2007), pp. 1433–1441. ISSN: 0890-6955. DOI: 10.1016/j.ijmachtools.2006.09.028.

[CHEN18] D. CHEN; X. ZHANG; Y. XIE, & H. DING: Precise Estimation of Cutting Force Coefficients and Cutter Runout in Milling using Differential Evolution Algorithm. In: *Procedia CIRP* 77 (2018), pp. 283–286. ISSN: 2212-8271. DOI: 10.1016/j.procir.2018.09.016.

[CHEN98] Z. CHEN; A. D. HOPE; H. SADEK, & G. SMITH: An LMI-based machining logic fusion. In: vol. 1998. IEE, 1998, pp. 519–524. ISBN: 978-0-85296-708-9. DOI: 10.1049/cp:19980283.

[CUS06] F. CUS; U. ZUPERL; E. KIKER, & M. MILFELNER: Adaptive controller design for feedrate maximization of machining process. In: *Journal of Achievements in Materials and Manufacturing Engineering*. Special Issue of AMME'2006 17.1 (Aug. 2006), pp. 237–240. URL: http://jamme.acmsse.h2.pl/papers_amme06/184.pdf.

[CUS11] F. CUS; U. ZUPERL, & J. BALIC: Combined feedforward and feedback control of end milling system. In: *Journal of Achievements in Materials and Manufacturing Engineering* 45.1 (2011), pp. 79–88. URL: http://jamme.acmsse.h2.pl/papers_vol45_1/45110.pdf.

[DANE86] L. K. DANESHMEND & H. A. PAK: Model Reference Adaptive Control of Feed Force in Turning. In: *Journal of Dynamic Systems, Measurement, and Control* 108.3 (Sept. 1, 1986), pp. 215–222. ISSN: 0022-0434. DOI: 10.1115/1.3143770.

[DENK12] B. DENKENA & F. FLÖTER: Adaptive Cutting Force Control on a Milling Machine with Hybrid Axis Configuration. In: *3rd CIRP Conference on Process Machine Interactions* 4 (2012), pp. 109–114. ISSN: 2212-8271. DOI: 10.1016/j.procir.2012.10.020.

[DHUP12] J. DHUPIA & I. GIRSANG: Correlation-based estimation of cutting force coefficients for ball-end milling. In: *Machining Science and Technology* 16.2 (Apr. 2012), pp. 287–303. ISSN: 1091-0344, 1532-2483. DOI: 10.1080/10910344.2012.673978.

[DOTC08] M. DOTCHEVA; H. MILLWARD, & A. LEWIS: The evaluation of cutting-force coefficients using surface error measurements. In: *Journal of Materials Processing Technology* 196.1 (Jan. 2008), pp. 42–51. ISSN: 0924-0136. DOI: 10.1016/j.jmatprotec.2007.04.136.

[DROS18] W.-G. DROSSEL; S. GEBHARDT; A. BUCHT; B. KRANZ; J. SCHNEIDER, & M. ETTRICHRÄTZ: Performance of a new piezoceramic thick film sensor for measurement and control of cutting forces during milling. In: *CIRP Annals* 67.1 (2018), pp. 45–48. ISSN: 00078506. DOI: 10.1016/j.cirp.2018.04.115.

[EHMA97] K. F. EHMANN; S. G. KAPOOR; R. E. DEVOR, & I. LAZOGLU: Machining Process Modeling: A Review. In: *Journal of Manu-*

facturing Science and Engineering 119.4 (1997), pp. 655–663. ISSN: 1087-1357. DOI: 10.1115/1.2836805.

[ENDR95] W. J. ENDRES; R. E. DEVOR, & S. G. KAPOOR: A Dual-Mechanism Approach to the Prediction of Machining Forces, Part 2: Calibration and Validation. In: *Journal of Engineering for Industry* 117.4 (1995), pp. 534–541. ISSN: 0022-0817. DOI: 10.1115/1.2803531.

[ERDI07] H. ERDIM; I. LAZOGLU, & M. KAYMAKCI: Free-form surface machining and comparing feedrate scheduling strategies. In: *Machining Science and Technology* 11.1 (Feb. 20, 2007), pp. 117–133. ISSN: 1091-0344, 1532-2483. DOI: 10.1080/10910340601172206.

[EVEN03] G. EVENSEN: The Ensemble Kalman Filter: theoretical formulation and practical implementation. In: *Ocean Dynamics* 53.4 (Nov. 1, 2003), pp. 343–367. ISSN: 1616-7341, 1616-7228. DOI: 10.1007/s10236-003-0036-9.

[EVEN94] G. EVENSEN: Sequential data assimilation with a nonlinear quasi-geostrophic model using Monte Carlo methods to forecast error statistics. In: *Journal of Geophysical Research* 99 (C5 1994), pp. 10143–10161. ISSN: 0148-0227. DOI: 10.1029/94JC00572.

[FERR17] H. J. FERREAU; A. POTSCHKA, & C. KIRCHES: *qpOASES*. Version 3.2.1. Apr. 26, 2017. URL: https://github.com/coin-or/qpOASES.

[FORB15] M. G. FORBES; R. S. PATWARDHAN; H. HAMADAH, & R. B. GOPALUNI: Model Predictive Control in Industry: Challenges and Opportunities. In: *IFAC-PapersOnLine* 48.8 (2015), pp. 531–538. ISSN: 2405-8963. DOI: 10.1016/j.ifacol.2015.09.022.

[FREI15] D. FREIBURG; R. HENSE; P. KERSTING, & D. BIERMANN: Determination of Force Parameters for Milling Simulations by Combining Optimization and Simulation Techniques. In: *Journal of Manufacturing Science and Engineering* 138.4 (Oct. 27, 2015), pp. 044502-1-044502–6. ISSN: 1087-1357. DOI: 10.1115/1.4031336.

[FRIE07] H. FRIEDIRCH: Über den Schnittwiderstand bei der Bearbeitung der Metalle durch Abheben von Spänen. In: *Zeitschrift des Vereins deutscher Ingenieure* 53.22 (1907), pp. 860–866.

[FU84] H. J. FU; R. E. DEVOR, & S. G. KAPOOR: A Mechanistic Model for the Prediction of the Force System in Face Milling Operations. In: *Journal of Engineering for Industry* 106.1 (1984), pp. 81–88. ISSN: 0022-0817. DOI: 10.1115/1.3185915.

[FUSS91] B. FUSSELL & K. SRINIVASAN: Adaptive control of force in end milling operations— an evaluation of available algorithms. In: *Journal of Manufacturing Systems* 10.1 (Jan. 1, 1991), pp. 8–20. ISSN: 0278-6125. DOI: 10.1016/0278-6125(91)90043-2.

[GAO13] G. GAO; B. WU; D. ZHANG, & M. LUO: Mechanistic identification of cutting force coefficients in bull-nose milling process. In: *Chinese Journal of Aeronautics* 26.3 (June 2013), pp. 823–830. ISSN: 1000-9361. DOI: 10.1016/j.cja.2013.04.007.

[GARC89] C. E. GARCÍA; D. M. PRETT, & M. MORARI: Model predictive control: Theory and practice—A survey. In: *Automatica* 25.3 (May 1989), pp. 335–348. ISSN: 0005-1098. DOI: 10.1016/0005-1098(89)90002-2.

[GARR10] J. L. GARRIGA & M. SOROUSH: Model Predictive Control Tuning Methods: A Review. In: *Industrial & Engineering Chemistry Research* 49.8 (Apr. 21, 2010), pp. 3505–3515. ISSN: 0888-5885, 1520-5045. DOI: 10.1021/ie900323c.

[GHOR15] H. GHORBANI & B. MOETAKEF-IMANI: Specific cutting force and cutting condition interaction modeling for round insert face milling operation. In: *The International Journal of Advanced Manufacturing Technology* (Oct. 24, 2015). ISSN: 0268-3768, 1433-3015. DOI: 10.1007/s00170-015-7985-2.

[GONZ10] O. GONZALO; J. BERISTAIN; H. JAUREGI, & C. SANZ: A method for the identification of the specific force coefficients for mechanistic milling simulation. In: *International Journal of Machine Tools and Manufacture* 50.9 (Sept. 2010), pp. 765–774. ISSN: 0890-6955. DOI: 10.1016/j.ijmachtools.2010.05.009.

[GRAD04] J. GRADIŠEK; M. KALVERAM, & K. WEINERT: Mechanistic identification of specific force coefficients for a general end mill. In: *International Journal of Machine Tools and Manufacture* 44.4 (Mar. 2004), pp. 401–414. ISSN: 0890-6955. DOI: 10.1016/j.ijmachtools.2003.10.001.

[GREW15] M. S. GREWAL & A. P. ANDREWS: *Kalman filtering: theory and practice using MATLAB*. Fourth edition. Hoboken, New Jersey: John Wiley & Sons Inc, 2015. 617 pp. ISBN: 978-1-118-85121-0.

[GROS17] N. GROSSI: Accurate and fast measurement of specific cutting force coefficients changing with spindle speed. In: *International Journal of Precision Engineering and Manufacturing* 18.8 (Aug. 2017), pp. 1173–1180. ISSN: 2234-7593, 2005-4602. DOI: 10.1007/s12541-017-0137-x.

[GUO17] Q. GUO; B. ZHAO; M. ZHANG; Y. JIANG, & Y. ZHANG: A separate-edge force coefficients' calibration method using specific condition for cutters with variable helix and pitch angles combining the runout effect. In: *The International Journal of Advanced Manufacturing Technology* 93.5 (Nov. 2017), pp. 1737–1749. ISSN: 0268-3768, 1433-3015. DOI: 10.1007/s00170-017-0628-z.

[GUO18] M. GUO; Z. WEI; M. WANG; S. LI, & S. LIU: An identification model of cutting force coefficients for five-axis ball-end milling. In: *The International Journal of Advanced Manufacturing Technology* 99.1 (Oct. 2018), pp. 937–949. ISSN: 0268-3768, 1433-3015. DOI: 10.1007/s00170-018-2451-6.

[HALL97] D. HALL & J. LLINAS: An introduction to multisensor data fusion. In: *Proceedings of the IEEE* 85.1 (Jan. 1997), pp. 6–23. ISSN: 0018-9219. DOI: 10.1109/5.554205.

[IBAR10] S. IBARAKI & T. SHIMIZU: A long-term control scheme of cutting forces to regulate tool life in end milling processes. In: *Precision Engineering* 34.4 (Oct. 2010), pp. 675–682. ISSN: 0141-6359. DOI: 10.1016/j.precisioneng.2010.05.001.

[IEEE-754] IEEE 754-2019: *IEEE Standard for Floating-Point Arithmetic*. Standard. Institute of Electrical & Electronics Engineers, July

2019. URL: https://standards.ieee.org/content/ieee-standards/en/standard/754-2019.html.

[IGLE13] M. A. IGLESIAS; K. J. H. LAW, & A. M. STUART: Ensemble Kalman methods for inverse problems. In: *Inverse Problems* 29.4 (Apr. 1, 2013), pp. 045001-1–045001-23. ISSN: 0266-5611, 1361-6420. DOI: 10.1088/0266-5611/29/4/045001.

[IJAZ14] F. IJAZ; N. MEHMOOD; M. JAMIL, & Y. AYAZ: Developments in adaptive control techniques for machine tools. In: *2014 International Conference on Robotics and Emerging Allied Technologies in Engineering (iCREATE)* (Apr. 22, 2014), pp. 186–190. DOI: 10.1109/iCREATE.2014.6828363.

[INTE97] INTERNATIONAL INSTITUTION FOR PRODUCTION ENGINEERING RESEARCH, ed.: *Dictionary of production engineering.* 2. edt. Berlin: Springer, 1997. 186 pp. ISBN: 978-3-540-20555-5.

[JAYA01] S. JAYARAM; S. KAPOOR, & R. DEVOR: Estimation of the specific cutting pressures for mechanistic cutting force models. In: *International Journal of Machine Tools and Manufacture* 41.2 (Jan. 1, 2001), pp. 265–281. ISSN: 0890-6955. DOI: 10.1016/S0890-6955(00)00076-6.

[JOHN20] S. G. JOHNSON: *The NLopt nonlinear-optimization package.* Version 2.6.2. Apr. 15, 2020. URL: http://github.com/stevengj/nlopt.

[JULI97] S. J. JULIER & J. K. UHLMANN: New extension of the Kalman filter to nonlinear systems. In: ed. by I. KADAR. Orlando, FL, USA, July 1997, p. 182. DOI: 10.1117/12.280797.

[KALM59] R. KALMAN: On the general theory of control systems. In: *IRE Transactions on Automatic Control* 4.3 (Dec. 1959), pp. 110–110. ISSN: 0096-199X. DOI: 10.1109/TAC.1959.1104873.

[KALM60] R. E. KALMAN: A New Approach to Linear Filtering and Prediction Problems. In: *Transactions of the ASME–Journal of Basic Engineering* 82 (Series D 1960), pp. 35–45. DOI: 10.1115/1.3662552.

[KAPO98] S. G. KAPOOR; R. E. DEVOR; R. ZHU; R. GAJJELA; G. PARAKKAL, & D. SMITHEY: Development of mechanistic models for the prediction of machining performance: model building methodology. In: *Machining Science and Technology* 2.2 (Dec. 1998), pp. 213–238. ISSN: 1091-0344, 1532-2483. DOI: 10.1080/10940349808945669.

[KAPP79] N. KAPPEN: *Entwicklung und Einsatz einer direkten digitalen Grenzregelung für eine Fräsmaschine mit CNC.* In collab. with UNIVERSITÄT STUTTGART. Berichte aus dem Institut für Steuerungstechnik der Werkzeugmaschinen und Fertigungseinrichtungen der Universität Stuttgart, ISW 22. Berlin: Springer, 1979. 123 pp. ISBN: 978-3-540-08591-1.

[KARA14] J. M. KARANDIKAR; T. L. SCHMITZ, & A. E. ABBAS: Application of Bayesian Inference to Milling Force Modeling. In: *Journal of Manufacturing Science and Engineering* 136.2 (Feb. 20, 2014), p. 12. ISSN: 1087-1357. DOI: 10.1115/1.4026365.

[KARU09] K. P. KARUNAKARAN; R. SHRINGI; D. RAMAMURTHI, & C. HARIHARAN: Octree-based NC simulation system for optimization of feed rate in milling using instantaneous force model. In: *The International Journal of Advanced Manufacturing Technology* 46.5 (June 2009), pp. 465–490. ISSN: 0268-3768, 1433-3015. DOI: 10.1007/s00170-009-2107-7.

[KELL14] D. T. B. KELLY; K. J. H. LAW, & A. M. STUART: Well-posedness and accuracy of the ensemble Kalman filter in discrete and continuous time. In: *Nonlinearity* 27.10 (Oct. 1, 2014), pp. 2579–2603. ISSN: 0951-7715, 1361-6544. DOI: 10.1088/0951-7715/27/10/2579.

[KIEN52] O. KIENZLE: Die Bestimmung von Kräften und Leistungen an spanenden Werkzeugmaschinen. In: *VDI-Z.* 81. VDI-Hauptversammlung der Fachgruppe Betriebstechnik. Vol. 96. VDI-Z. VDI-Verlag, 1952, pp. 299–305.

[KIEN57] O. KIENZLE & H. VICTOR: Spezifische Schnittkräfte bei der Metallbearbeitung. DE. In: *Werkstattstechnik und Maschinenbau* 47 (1957), pp. 224–225.

[KIM96] T.-Y. KIM & J. KIM: Adaptive cutting force control for a machining center by using indirect cutting force measurements. In: *International Journal of Machine Tools and Manufacture* 36.8

(Aug. 1996), pp. 925–937. ISSN: 0890-6955. DOI: 10.1016/0890-6955(96)00097-1.

[KLIN82] W. A. KLINE: The prediction of cutting forces and surface accuracy for the end milling process. Dr.-Phil. Urbana, Illinois: University of Illinois at Urbana-Champaign, 1982. 208 pp.

[KLIN83] W. KLINE & R. DEVOR: The effect of runout on cutting geometry and forces in end milling. In: *International Journal of Machine Tool Design and Research* 23.2 (Jan. 1983), pp. 123–140. ISSN: 0020-7357. DOI: 10.1016/0020-7357(83)90012-4.

[KLUG99] R. KLUGE: *Prädiktive modellbasierte Regelung von Zerspanprozessen*. Red. by G. WARNEKE. Vol. 35. FBK Produktionstechnische Berichte. Kaiserslautern: Universität Kaiserslautern, Aug. 17, 1999. 127 pp.

[KÖNI82] W. KÖNIG; K. ESSEL; L. WITTE, & VEREIN DEUTSCHER EISENHÜTTENLEUTE: *Spezifische Schnittkraftwerte für die Zerspanung metallischer Werkstoffe*. Düsseldorf: Verlag Stahleisen, 1982. ISBN: 978-3-514-00129-9.

[KORE81] Y. KOREN; O. MASORY, & S. MALKIN: Adaptive Control with Process Estimation. In: *CIRP Annals - Manufacturing Technology* 30.1 (1981), pp. 373–376. ISSN: 0007-8506. DOI: 10.1016/S0007-8506(07)60960-3.

[KRÜG13] M. KRÜGER & B. DENKENA: Model-based identification of tool runout in end milling and estimation of surface roughness from measured cutting forces. In: *The International Journal of Advanced Manufacturing Technology* 65.5 (Mar. 2013), pp. 1067–1080. ISSN: 0268-3768, 1433-3015. DOI: 10.1007/s00170-012-4240-y.

[KURT11] M. KURT & E. BAGCI: Feedrate optimisation/scheduling on sculptured surface machining: a comprehensive review, applications and future directions. In: *The International Journal of Advanced Manufacturing Technology* 55.9 (Apr. 2011), pp. 1037–1067. ISSN: 0268-3768, 1433-3015. DOI: 10.1007/s00170-010-3131-3.

[LAUD89] L. K. LAUDERBAUGH & A. G. ULSOY: Model Reference Adaptive Force Control in Milling. In: *Journal of Engineering for Industry* 111.1 (Feb. 1, 1989), pp. 13–21. ISSN: 1087-1357. DOI: 10.1115/1.3188726.

[LAYE12] K. S. E. LAYEGH; E. HUSEYIN, & I. LAZOGLU: Offline Force Control and Feedrate Scheduling for Complex Free Form Surfaces in 5-Axis Milling. In: *Procedia CIRP* 1 (2012), pp. 96–101. ISSN: 2212-8271. DOI: 10.1016/j.procir.2012.04.015.

[LI17] Z.-L. LI; Y. DING, & L.-M. ZHU: Accurate cutting force prediction of helical milling operations considering the cutter runout effect. In: *The International Journal of Advanced Manufacturing Technology* 92.9 (Oct. 2017), pp. 4133–4144. ISSN: 0268-3768, 1433-3015. DOI: 10.1007/s00170-017-0464-1.

[LIAN04] S. Y. LIANG; R. L. HECKER, & R. G. LANDERS: Machining Process Monitoring and Control: The State-of-the-Art. In: *Journal of Manufacturing Science and Engineering* 126.2 (May 1, 2004), pp. 297–310. ISSN: 1087-1357, 1528-8935. DOI: 10.1115/1.1707035.

[LIU18] X. LIU; H. GAO; C. YUE; R. LI; N. JIANG, & L. YANG: Investigation of the milling stability based on modified variable cutting force coefficients. In: *The International Journal of Advanced Manufacturing Technology* 96.9 (June 2018), pp. 2991–3002. ISSN: 0268-3768, 1433-3015. DOI: 10.1007/s00170-018-1780-9.

[LIU19] D. LIU; Y. HU; D. ZHANG, & H. LUO: Milling force monitoring with thin-film sensors integrated into fixtures. In: *The International Journal of Advanced Manufacturing Technology* 103.1 (July 2019), pp. 1519–1527. ISSN: 0268-3768, 1433-3015. DOI: 10.1007/s00170-019-03666-6.

[LUO18a] M. LUO; Z. CHONG, & D. LIU: Cutting Forces Measurement for Milling Process by Using Working Tables with Integrated PVDF Thin-Film Sensors. In: *Sensors* 18.11 (Nov. 19, 2018), pp. 4031-1–4031-13. ISSN: 1424-8220. DOI: 10.3390/s18114031.

[LUO18b] M. LUO; H. LUO; D. AXINTE; D. LIU; J. MEI, & Z. LIAO: A wireless instrumented milling cutter system with embedded

PVDF sensors. In: *Mechanical Systems and Signal Processing* 110 (Sept. 2018), pp. 556–568. ISSN: 0888-3270. DOI: 10.1016/j.ymssp.2018.03.040.

[LUTT98] C. van LUTTERVELT; T. CHILDS; I. JAWAHIR; F. KLOCKE; P. VENUVINOD; Y. ALTINTAS; E. ARMAREGO; D. DORNFELD; I. GRABEC; J. LEOPOLD; B. LINDSTROM; D. LUCCA; T. OBIKAWA; SHIRAKASHI, & H. SATO: Present Situation and Future Trends in Modelling of Machining Operations Progress Report of the CIRP Working Group 'Modelling of Machining Operations'. In: *CIRP Annals* 47.2 (1998), pp. 587–626. ISSN: 0007-8506. DOI: 10.1016/S0007-8506(07)63244-2.

[MA12] L. MA; S. N. MELKOTE; J. B. MOREHOUSE; J. B. CASTLE; J. W. FONDA, & M. A. JOHNSON: Thin-Film PVDF Sensor-Based Monitoring of Cutting Forces in Peripheral End Milling. In: *Journal of Dynamic Systems, Measurement, and Control* 134.5 (Sept. 1, 2012), p. 051014. ISSN: 0022-0434, 1528-9028. DOI: 10.1115/1.4006366.

[MA19] H. MA; W. LIU; X. ZHOU; Q. NIU, & C. KONG: An effective and automatic approach for parameters optimization of complex end milling process based on virtual machining. In: *Journal of Intelligent Manufacturing* (Sept. 11, 2019). ISSN: 0956-5515, 1572-8145. DOI: 10.1007/s10845-019-01489-6.

[MARP21] MARPOSS S.P.A.: *Adaptive Control for Monitoring Systems*. 2021. URL: https://www.marposs.com/eng/product/adaptive-control-for-monitoring-systems (visited on 01/03/2021).

[MART41] M. E. MARTELLOTTI: An analysis of the milling process. In: *Transactions of the ASME* 63 (1941), pp. 677–700.

[MATS09] A. MATSUBARA & S. IBARAKI: Monitoring and Control of Cutting Forces in Machining Processes: A Review. In: *International Journal of Automation Technology* 3.4 (July 5, 2009), pp. 445–456. ISSN: 1883-8022, 1881-7629. DOI: 10.20965/ijat.2009.p0445.

[MATS17] T. MATSUMURA & S. TAMURA: Cutting Force Model in Milling with Cutter Runout. In: *Procedia CIRP* 58 (2017), pp. 566–571. ISSN: 2212-8271. DOI: 10.1016/j.procir.2017.03.268.

[MAYB79] P. S. MAYBECK: *Stochastic models, estimation and control.* Mathematics in science and engineering v. 141. New York: Academic Press, 1979. ISBN: 978-0-08-095650-3.

[MAYN00] D. MAYNE; J. RAWLINGS; C. RAO, & P. SCOKAERT: Constrained model predictive control: Stability and optimality. In: *Automatica* 36.6 (June 2000), pp. 789–814. ISSN: 0005-1098. DOI: 10.1016/S0005-1098(99)00214-9.

[MCU 21] MCU GMBH & CO. KG: *Prozessüberwachungssystem Toolinspect.* 2021. URL: https://www.mcu-gmbh.de/produkte/toolinspect (visited on 01/03/2021).

[MIUR18] K. MIURA; B. DÖBBELER, & F. KLOCKE: Cutting Power Estimation via External Voltage and Current Sensors on Feed-Drive Axis for the Straight Milling Process. In: *Procedia CIRP* 78 (2018), pp. 323–328. ISSN: 2212-8271. DOI: 10.1016/j.procir.2018.09.068.

[MIUR19] K. MIURA & T. BERGS: A Method of Cutting Power Monitoring for Feed Axes in Milling by Power Measurement Device. In: *IFAC-PapersOnLine* 52.13 (2019), pp. 2471–2476. ISSN: 2405-8963. DOI: 10.1016/j.ifacol.2019.11.577.

[MODU21] MODULEWORKS GMBH: Fact Sheet: Adaptive Roughing Component. EN. 2021. URL: https://www.moduleworks.com/?jet_download=10818 (visited on 01/03/2021).

[NELS76] L. NELSON & E. STEAR: The simultaneous on-line estimation of parameters and states in linear systems. In: *IEEE Transactions on Automatic Control* 21.1 (Feb. 1976), pp. 94–98. ISSN: 0018-9286. DOI: 10.1109/TAC.1976.1101148.

[NOLZ95a] H. NOLZEN & R. ISERMANN: Adaptive control of the cutting power for milling operation. In: *American Control Conference, Proceedings of the 1995* 3 (June 21, 1995), 2185–2189 vol.3. DOI: 10.1109/ACC.1995.531287.

[NOLZ95b] H. NOLZEN & R. ISERMANN: Fast adaptive cutting force control for milling operation. In: *Control Applications, 1995., Proceedings of the 4th IEEE Conference on* (Sept. 28, 1995), pp. 760–765. DOI: 10.1109/CCA.1995.555844.

[NOLZ97] H. NOLZEN: *Parameteradaptive Regelung von Fräsprozessen.* Als Ms. gedr. Fortschritt-Berichte VDI Reihe 8, Meß-, Steuerungs- und Regelungstechnik 623. Düsseldorf: VDI-Verl, 1997. 153 pp. ISBN: 978-3-18-362308-2.

[NOUR15] M. NOURI; B. K. FUSSELL; B. L. ZINITI, & E. LINDER: Real-time tool wear monitoring in milling using a cutting condition independent method. In: *International Journal of Machine Tools and Manufacture* 89 (Feb. 2015), pp. 1–13. ISSN: 0890-6955. DOI: 10.1016/j.ijmachtools.2014.10.011.

[PARK03] S. S. PARK: High frequency bandwidth cutting force measurements in milling using the spindle integrated force sensor system. PhD thesis. Vancouver, Canada: University of British Columbia, 2003. 249 pp. DOI: 10.14288/1.0081000.

[PARK04] S. S. PARK & Y. ALTINTAS: Dynamic Compensation of Spindle Integrated Force Sensors With Kalman Filter. In: *Journal of Dynamic Systems, Measurement, and Control* 126.3 (2004), pp. 443–452. ISSN: 0022-0434. DOI: 10.1115/1.1789531.

[PARK18] H.-S. PARK; B. QI; D.-V. DANG, & D. Y. PARK: Development of smart machining system for optimizing feedrates to minimize machining time. In: *Journal of Computational Design and Engineering* 5.3 (July 1, 2018), pp. 299–304. ISSN: 2288-5048. DOI: 10.1016/j.jcde.2017.12.004.

[PERE13] H. PEREZ; E. DIEZ; J. J. MARQUEZ, & A. VIZAN: An enhanced method for cutting force estimation in peripheral milling. In: *The International Journal of Advanced Manufacturing Technology* 69.5 (Nov. 2013), pp. 1731–1741. ISSN: 0268-3768, 1433-3015. DOI: 10.1007/s00170-013-5153-0.

[PLET18] A. PLETA; F. A. NIAKI, & L. MEARS: A comparative study on the cutting force coefficient identification between trochoidal and slot milling. In: *Procedia Manufacturing* 26 (2018), pp. 570–579. ISSN: 2351-9789. DOI: 10.1016/j.promfg.2018.07.067.

[POST19] M. POSTEL; D. ASLAN; K. WEGENER, & Y. ALTINTAS: Monitoring of vibrations and cutting forces with spindle mounted vibration sensors. In: *CIRP Annals* 68.1 (2019), pp. 413–416. ISSN: 0007-8506. DOI: 10.1016/j.cirp.2019.03.019.

[POWE04] M. POWELL: The NEWUOA software for unconstrained optimization without derivatives. Nov. 2004. URL: http://www.damtp.cam.ac.uk/user/na/NA_papers/NA2004_08.pdf.

[POWE07] M. POWELL: Developments of NEWUOA for unconstrained minimization without derivatives. 2007. URL: http://www.damtp.cam.ac.uk/user/na/NA_papers/NA2007_05.pdf.

[POWE09] M. POWELL: The BOBYQA algorithm for bound constrained optimization without derivatives. Aug. 2009. URL: http://www.damtp.cam.ac.uk/user/na/NA_papers/NA2009_06.pdf.

[RAWL00] J. B. RAWLINGS: Tutorial overview of model predictive control. In: *IEEE Control Systems* 20.3 (June 2000), pp. 38–52. ISSN: 1066-033X. DOI: 10.1109/37.845037.

[REKE19] S. REKERS: *Correction of Systematic Errors in Piezoelectric Cutting Force Measurement.* Aachen , Germany: Apprimus Verlag, 2019. ISBN: 978-3-86359-747-4.

[RICH02] N. D. RICHARDS; B. K. FUSSELL, & R. B. JERARD: Efficient NC Machining Using Off-Line Optimized Feedrates and On-Line Adaptive Control. In: 3641 (2002), pp. 181–191. DOI: 10.1115/IMECE2002-33618.

[RICH78] J. RICHALET; A. RAULT; J. TESTUD, & J. PAPON: Model predictive heuristic control: Applications to industrial processes. In: *Automatica* 14.5 (Sept. 1, 1978), pp. 413–428. ISSN: 0005-1098. DOI: 10.1016/0005-1098(78)90001-8.

[RICH93] J. RICHALET: Industrial applications of model based predictive control. In: *Automatica* 29.5 (Sept. 1993), pp. 1251–1274. ISSN: 0005-1098. DOI: 10.1016/0005-1098(93)90049-Y.

[RIVI09] E. RIVIÈRE-LORPHÈVRE & E. FILIPPI: Mechanistic cutting force model parameters evaluation in milling taking cutter radial runout into account. In: *The International Journal of Advanced Manufacturing Technology* 45.1 (Nov. 2009), pp. 8–15. ISSN: 0268-3768, 1433-3015. DOI: 10.1007/s00170-009-1943-9.

[ROBE96] S. J. ROBER & Y. C. SHIN: Control of Cutting Force for End Milling Processes Using an Extended Model Reference Adaptive Control Scheme. In: *Journal of Manufacturing Science and*

Engineering 118.3 (Aug. 1, 1996), pp. 339–347. ISSN: 1087-1357. DOI: 10.1115/1.2831035.

[ROBE97] S. J. ROBER; Y. C. SHIN, & O. D. I. NWOKAH: A Digital Robust Controller for Cutting Force Control in the End Milling Process. In: *Journal of Dynamic Systems, Measurement, and Control* 119.2 (June 1, 1997), pp. 146–152. ISSN: 0022-0434, 1528-9028. DOI: 10.1115/1.2801226.

[RUBE16a] M. A. RUBEO & T. L. SCHMITZ: Mechanistic force model coefficients: A comparison of linear regression and nonlinear optimization. In: *Precision Engineering* 45 (July 2016), pp. 311–321. ISSN: 0141-6359. DOI: 10.1016/j.precisioneng.2016.03.008.

[RUBE16b] M. A. RUBEO & T. L. SCHMITZ: Milling Force Modeling: A Comparison of Two Approaches. In: *Procedia Manufacturing* 5 (2016), pp. 90–105. ISSN: 2351-9789. DOI: 10.1016/j.promfg.2016.08.010.

[RUBI13] L. RUBIO; A. LONGSTAFF; S. FLETCHER; A. MYERS, & M. DE LA SEN: Control Scheme for Automatically Controlling the Milling System in between Optimal Working Points. In: 2013 IEEE International Conference on Systems, Man, and Cybernetics (SMC 2013). Manchester: IEEE, Oct. 2013, pp. 1628–1633. ISBN: 978-1-4799-0652-9. DOI: 10.1109/SMC.2013.281.

[SANZ19] D. SANZ-ALONSO; A. M. STUART, & A. TAEB: Inverse Problems and Data Assimilation. In: *arXiv:1810.06191* (July 3, 2019). (Visited on 09/10/2020).

[SCHE79] L. SCHENKE: *Auslegung einer technologisch-geometrischen Grenzregelung für die Fräsbearbeitung.* In collab. with UNIVERSITÄT STUTTGART. Berichte aus dem Institut für Steuerungstechnik der Werkzeugmaschinen und Fertigungseinrichtungen der Universität Stuttgart, ISW 26. Berlin: Springer, 1979. 113 pp. ISBN: 978-3-540-09222-3.

[SCHI17] C. SCHILLINGS & A. M. STUART: Analysis of the Ensemble Kalman Filter for Inverse Problems. In: *SIAM Journal on Numerical Analysis* 55.3 (Jan. 2017), pp. 1264–1290. ISSN: 0036-1429, 1095-7170. DOI: 10.1137/16M105959X.

[SCHI18] C. SCHILLINGS & A. M. STUART: Convergence analysis of ensemble Kalman inversion: the linear, noisy case. In: *Applicable Analysis* 97.1 (Jan. 2, 2018), pp. 107–123. ISSN: 0003-6811, 1563-504X. DOI: 10.1080/00036811.2017.1386784.

[SCHM07] T. L. SCHMITZ; J. COUEY; E. MARSH; N. MAUNTLER, & D. HUGHES: Runout effects in milling: Surface finish, surface location error, and stability. In: *International Journal of Machine Tools and Manufacture* 47.5 (Apr. 2007), pp. 841–851. ISSN: 0890-6955. DOI: 10.1016/j.ijmachtools.2006.06.014.

[SCHR77] K.-W. SCHRICK, ed.: *Anwendungen der Kalman-Filter-Technik: Anleitung und Beispiele.* Methoden der Regelungstechnik. München: Oldenbourg, 1977. 353 pp. ISBN: 978-3-486-20931-0.

[SCHW17] M. SCHWENZER; O. ADAMS; F. KLOCKE; S. STEMMLER, & D. ABEL: Model-based predictive force control in milling: determination of reference trajectory. In: *Production Engineering* 11.2 (Apr. 2017), pp. 107–115. ISSN: 0944-6524, 1863-7353. DOI: 10.1007/s11740-017-0721-z.

[SCHW19a] M. SCHWENZER; T. AUERBACH; B. DÖBBELER, & T. BERGS: Comparative study on optimization algorithms for online identification of an instantaneous force model in milling. In: *The International Journal of Advanced Manufacturing Technology* 101.9 (Apr. 2019), pp. 2249–2257. ISSN: 0268-3768, 1433-3015. DOI: 10.1007/s00170-018-3109-0.

[SCHW19b] M. SCHWENZER; S. STEMMLER; M. AY; T. BERGS, & D. ABEL: Continuous identification for mechanistic force models in milling. In: *IFAC-PapersOnLine* 52.13 (2019), pp. 1791–1796. ISSN: 2405-8963. DOI: 10.1016/j.ifacol.2019.11.461.

[SCHW20] M. SCHWENZER; G. VISCONTI; M. AY; T. BERGS; M. HERTY, & D. ABEL: Identifying trending model coefficients with an ensemble Kalman filter – a demonstration on a force model for milling. EN. In: *IFAC-PapersOnLine* 53.2 (2020), pp. 2292–2298. ISSN: 24058963. DOI: 10.1016/j.ifacol.2020.12.1490.

[SCHW21] M. SCHWENZER; M. AY; T. BERGS, & D. ABEL: Review on model predictive control: an engineering perspective. EN. In: *The International Journal of Advanced Manufacturing Technology* (Aug.

2021). ISSN: 0268-3768, 1433-3015. DOI: 10.1007/s00170-021-07682-3.

[SEET99] R. J. SEETHALER & I. YELLOWLEY: The Identification of Radial Runout in Milling Operations. In: *Journal of Manufacturing Science and Engineering* 121.3 (1999), pp. 524–531. ISSN: 1087-1357. DOI: 10.1115/1.2832712.

[SHIN97] Y. C. SHIN & A. J. WATERS: A new procedure to determine instantaneous cutting force coefficients for machining force prediction. In: *International Journal of Machine Tools and Manufacture* 37.9 (Sept. 1997), pp. 1337–1351. ISSN: 0890-6955. DOI: 10.1016/S0890-6955(96)00093-4.

[SIEM16] SIEMENS INDUSTRY ONLINE SUPPORT, ed.: Guide for Commissioning a Position-Controlled Drive. Mar. 7, 2016. URL: https://support.industry.siemens.com/cs/ww/en/view/109479977 (visited on 03/07/2019).

[SIMO06] D. SIMON: *Optimal state estimation: Kalman, H [infinity] and nonlinear approaches*. Hoboken, N.J: Wiley-Interscience, 2006. 526 pp. ISBN: 978-0-471-70858-2.

[SOLI21] SOLIDCAM GMBH: *iMachining - the Revolution in CNC Machining*. 2021. URL: https://www.solidcam.com/en-us/imachining-the-revolution-in-cnc-machining/ (visited on 01/03/2021).

[SPEN91] A. SPENCE & Y. ALTINTAS: CAD Assisted Adaptive Control for Milling. In: *Journal of Dynamic Systems, Measurement, and Control* 113.3 (Sept. 1, 1991), pp. 444–450. ISSN: 0022-0434. DOI: 10.1115/1.2896430.

[STEM16] S. STEMMLER; D. ABEL; O. ADAMS, & F. KLOCKE: Model Predictive Feed Rate Control for a Milling Machine. In: *8th IFAC Conference on Manufacturing Modelling, Management and Control MIM 2016Troyes, France, 28—30 June 2016* 49.12 (2016), pp. 11–16. ISSN: 2405-8963. DOI: 10.1016/j.ifacol.2016.07.542.

[STEM19] S. STEMMLER; M. AY; M. SCHWENZER; D. ABEL, & T. BERGS: Model-based Predictive Force Control in Milling. In: 2019 18th European Control Conference (ECC). Naples, Italy: IEEE, June

2019, pp. 4313–4318. ISBN: 978-3-907144-00-8. DOI: 10.23919/ECC.2019.8795716.

[STEM20] S. STEMMLER: Intelligente Regelungsstrategien als Schlüsseltechnologie selbstoptimierender Fertigungssysteme. Dissertation. Aachen: RWTH Aachen University, 2020. 205 pp. DOI: 10.18154/RWTH-2020-02766.

[STEP11] M. A. STEPHENS; C. MANZIE, & M. C. GOOD: Explicit Model Predictive Control for Reference Tracking on an Industrial Machine Tool. In: *IFAC Proceedings Volumes* 44.1 (Jan. 2011), pp. 14513–14518. ISSN: 1474-6670. DOI: 10.3182/20110828-6-IT-1002.00579.

[STEP30] B. STEPHEN: On the Theory of Filter Amplifiers. In: *Experimental Wireless and the Wireless Engineer,* 7 (1930), pp. 536–541.

[STRE79] V. STREJC: Least Squares Parameter Estimation. In: *IFAC Proceedings Volumes* 12.8 (Sept. 1979), pp. 535–550. ISSN: 1474-6670. DOI: 10.1016/S1474-6670(17)53975-0.

[STUT76] G. STUTE & F. R. GOETZ: Adaptive Control System for Variable Gain in Acc Systems. In: *Proceedings of the Sixteenth International Machine Tool Design and Research Conference.* Ed. by F. KOENIGSBERGER & S. A. TOBIAS. London: Macmillan Education UK, 1976, pp. 117–121. ISBN: 978-1-349-81544-9. DOI: 10.1007/978-1-349-81544-9_18.

[SUTH86] J. W. SUTHERLAND & R. E. DEVOR: An Improved Method for Cutting Force and Surface Error Prediction in Flexible End Milling Systems. In: *Journal of Engineering for Industry* 108.4 (1986), p. 269. ISSN: 0022-0817. DOI: 10.1115/1.3187077.

[TAYL07] F. W. TAYLOR: *On the art of cutting metals.* New York: The american society of mechanical engineers, 1907. 565 pp.

[THOM18] U. THOMBANSEN; G. BUCHHOLZ; D. FRANK; J. HEINISCH; M. KEMPER; T. PULLEN; V. REIMER; G. ROTSHTEYN; M. SCHWENZER; S. STEMMLER; D. ABEL; T. GRIES; C. HOPMANN; F. KLOCKE; R. POPRAWE; U. REISGEN, & R. SCHMITT: Design framework for model-based self-optimizing manufacturing systems. In: *The International Journal of Advanced Manufacturing*

Technology 97.1 (July 2018), pp. 519–528. ISSN: 0268-3768, 1433-3015. DOI: 10.1007/s00170-018-1951-8.

[TSAI16] M. TSAI; S. CHANG; J. HUNG, & C. WANG: Investigation of milling cutting forces and cutting coefficient for aluminum 6060-T6. In: *Computers & Electrical Engineering* 51 (Apr. 2016), pp. 320–330. ISSN: 00457-906. DOI: 10.1016/j.compeleceng.2015.09.016.

[TU04] J. F. TU; M. CORLESS, & J. JEPPSSON: Robust Control of High Speed End Milling with Unknown Process Parameter and CNC Delay. In: *Journal of Aerospace Engineering* 17.1 (Jan. 2004), pp. 1–9. ISSN: 0893-1321, 1943-5525. DOI: 10.1061/(ASCE)0893-1321(2004)17:1(1).

[TU14] J. F. TU & M. CORLESS: Review of Sensor-Based Approach to Reliable High Speed Machining at Boeing – A Tribute to Jan Jeppsson. In: *High Speed Machining* 1.1 (Jan. 15, 2014). ISSN: 2299-3975. DOI: 10.2478/hsm-2014-0001.

[TUKO11] B. TUKORA & T. SZALAY: Real-time determination of cutting force coefficients without cutting geometry restriction. In: *International Journal of Machine Tools and Manufacture* 51.12 (Dec. 2011), pp. 871–879. ISSN: 0890-6955. DOI: 10.1016/j.ijmachtools.2011.08.003.

[ULSO89] A. G. ULSOY & Y. KOREN: Applications of adaptive control to machine tool process control. In: *IEEE Control Systems Magazine* 9.4 (June 1989), pp. 33–37. ISSN: 0272-1708. DOI: 10.1109/37.24837.

[ULSO93] A. G. ULSOY & Y. KOREN: Control of Machining Processes. In: *Journal of Dynamic Systems, Measurement, and Control* 115.2 (June 1, 1993), pp. 301–308. ISSN: 0022-0434, 1528-9028. DOI: 10.1115/1.2899070.

[VICT56] H. VICTOR: Beitrag zur Kenntnis der Schnittgräfte beim Drehen, Hobeln und Bohren. DE. Dr.-Ing. Hannover: TH Hannover, 1956.

[WALD99] D. J. WALDORF; R. E. DEVOR, & S. G. KAPOOR: An Evaluation of Ploughing Models for Orthogonal Machining. In: *Journal of*

Manufacturing Science and Engineering 121.4 (1999), p. 550. ISSN: 1087-1357. DOI: 10.1115/1.2833050.

[WAN01] E. A. WAN & A. T. NELSON: Dual Extended Kalman Filter Methods. In: *Kalman Filtering and Neural Networks*. Ed. by S. HAYKIN. New York, USA: John Wiley & Sons, Inc., Oct. 1, 2001, pp. 123–173. ISBN: 978-0-471-22154-8. DOI: 10.1002/0471221546.ch5.

[WAN07a] M. WAN; W. H. ZHANG; G. TAN, & G. H. QIN: New algorithm for calibration of instantaneous cutting-force coefficients and radial run-out parameters in flat end milling. In: *Proceedings of the Institution of Mechanical Engineers, Part B: Journal of Engineering Manufacture* 221.6 (June 2007), pp. 1007–1019. ISSN: 0954-4054, 2041-2975. DOI: 10.1243/09544054JEM515.

[WAN07b] M. WAN; W. ZHANG; G. QIN, & G. TAN: Efficient calibration of instantaneous cutting force coefficients and runout parameters for general end mills. In: *International Journal of Machine Tools and Manufacture* 47.11 (Sept. 2007), pp. 1767–1776. ISSN: 0890-6955. DOI: 10.1016/j.ijmachtools.2006.06.012.

[WAN08] M. WAN; W. H. ZHANG; G. H. QIN, & Z. P. WANG: Consistency study on three cutting force modelling methods for peripheral milling. In: *Proceedings of the Institution of Mechanical Engineers, Part B: Journal of Engineering Manufacture* 222.6 (June 2008), pp. 665–676. ISSN: 0954-4054, 2041-2975. DOI: 10.1243/09544054JEM1085.

[WAN09a] M. WAN & W.-H. ZHANG: Systematic study on cutting force modelling methods for peripheral milling. In: *International Journal of Machine Tools and Manufacture* 49.5 (Apr. 2009), pp. 424–432. ISSN: 0890-6955. DOI: 10.1016/j.ijmachtools.2008.12.004.

[WAN09b] M. WAN; W.-H. ZHANG; J.-W. DANG, & Y. YANG: New procedures for calibration of instantaneous cutting force coefficients and cutter runout parameters in peripheral milling. In: *International Journal of Machine Tools and Manufacture* 49.14 (Nov. 2009), pp. 1144–1151. ISSN: 0890-6955. DOI: 10.1016/j.ijmachtools.2009.08.005.

[WANG02] J. J. J. WANG & C. ZHENG: Identification of shearing and ploughing cutting constants from average forces in ball-end milling. In: *International Journal of Machine Tools and Manufacture* 42.6 (May 2002), pp. 695–705. ISSN: 0890-6955. DOI: 10.1016/S0890-6955(02)00002-0.

[WANG03a] J. J. J. WANG & C. M. ZHENG: Online Identification of Shearing and Plowing Constants in End Milling. In: *Journal of Manufacturing Science and Engineering* 125.1 (2003), pp. 57–64. ISSN: 1087-1357. DOI: 10.1115/1.1536931.

[WANG03b] J. J. J. WANG & C. ZHENG: Identification of cutter offset in end milling without a prior knowledge of cutting coefficients. In: *International Journal of Machine Tools and Manufacture* 43.7 (May 2003), pp. 687–697. ISSN: 0890-6955. DOI: 10.1016/S0890-6955(03)00028-2.

[WANG04] J. J. J. WANG & H. C. CHANG: Extracting cutting constants via harmonic force components for a general helical end mill. In: *The International Journal of Advanced Manufacturing Technology* 24.5 (Sept. 2004), pp. 415–424. ISSN: 0268-3768, 1433-3015. DOI: 10.1007/s00170-003-1664-4.

[WANG12] H. WANG; X. QIN; C. REN, & Q. WANG: Prediction of cutting forces in helical milling process. In: *The International Journal of Advanced Manufacturing Technology* 58.9 (Feb. 2012), pp. 849–859. ISSN: 0268-3768, 1433-3015. DOI: 10.1007/s00170-011-3435-y.

[WANG13] B. WANG; H. HAO; M. WANG; J. HOU, & Y. FENG: Identification of instantaneous cutting force coefficients using surface error. In: *The International Journal of Advanced Manufacturing Technology* 68.1 (Sept. 2013), pp. 701–709. ISSN: 0268-3768, 1433-3015. DOI: 10.1007/s00170-013-4792-5.

[WANG14] M. WANG; L. GAO, & Y. ZHENG: An examination of the fundamental mechanics of cutting force coefficients. In: *International Journal of Machine Tools and Manufacture* 78 (Mar. 2014), pp. 1–7. ISSN: 0890-6955. DOI: 10.1016/j.ijmachtools.2013.10.008.

[WANG18] L. WANG; H. SI; L. GUAN, & Z. LIU: Comparison of different polynomial functions for predicting cutting coefficients in the milling process. In: *The International Journal of Advanced Manufacturing Technology* 94.5 (Feb. 2018), pp. 2961–2972. ISSN: 0268-3768, 1433-3015. DOI: 10.1007/s00170-017-1086-3.

[WECK06] M. WECK & C. BRECHER: *Mechatronische Systeme, Vorschubantriebe, Prozessdiagnose*. 6., rev. edt. Werkzeugmaschinen, Fertigungssysteme Manfred Weck; Christian Brecher; 3. Berlin: Springer, 2006. 424 pp. ISBN: 978-3-540-22506-5.

[WEI18] Z. C. WEI; M. L. GUO; M. J. WANG; S. Q. LI, & S. X. LIU: Prediction of cutting force in five-axis flat-end milling. In: *The International Journal of Advanced Manufacturing Technology* (Jan. 19, 2018). ISSN: 0268-3768, 1433-3015. DOI: 10.1007/s00170-017-1380-0.

[WIMM18] S. WIMMER; J. ELLINGER, & M. F. ZAEH: A cutting force model for finishing processes using helical end mills with significant runout. In: *Production Engineering* 12.6 (Dec. 2018), pp. 703–714. ISSN: 0944-6524, 1863-7353. DOI: 10.1007/s11740-018-0846-8.

[WIRT10] G. F. WIRTZ: *Adaptive Regelung für das Tiefbohren mit Einlippenbohrern kleiner Durchmesser*. 1. edt. Ergebnisse aus der Produktionstechnik Technologie der Fertigungsverfahren 2010,9. Aachen: Apprimus-Verl, 2010. 118 pp. ISBN: 978-3-940565-56-3.

[WITT19] M. WITT; M. SCHUMANN, & P. KLIMANT: Real-time machine simulation using cutting force calculation based on a voxel material removal model. In: *The International Journal of Advanced Manufacturing Technology* 105.5 (Dec. 2019), pp. 2321–2328. ISSN: 0268-3768, 1433-3015. DOI: 10.1007/s00170-019-04418-2.

[WOJC15] S. WOJCIECHOWSKI: The estimation of cutting forces and specific force coefficients during finishing ball end milling of inclined surfaces. In: *International Journal of Machine Tools and Manufacture* 89 (Feb. 2015), pp. 110–123. ISSN: 0890-6955. DOI: 10.1016/j.ijmachtools.2014.10.006.

[XU08] C. XU & Y. C. SHIN: An Adaptive Fuzzy Controller for Constant Cutting Force in End-Milling Processes. In: *Journal of Manufacturing Science and Engineering* 130.3 (2008), pp. 031001-1–031001-10. ISSN: 1087-1357. DOI: 10.1115/1.2823070.

[XU96] G. XU: *Einfluß der Schneidkantenform auf die Oberflächenausbildung beim Hochgeschwindigkeitsfräsen mit Feinkornhartmetall.* Als Ms. gedr. Darmstädter Forschungsberichte für Konstruktion und Fertigung. Aachen: Shaker, 1996. 131 pp. ISBN: 978-3-8265-1606-1.

[YAMA91] K. YAMAZAKI; N. KOJIMA; C. SAKAMOTO, & T. SAITO: Real-Time Model Reference Adaptive Control of 3-D Sculptured Surface Machining. In: *CIRP Annals - Manufacturing Technology* 40.1 (Jan. 1, 1991), pp. 479–482. ISSN: 0007-8506. DOI: 10.1016/S0007-8506(07)62034-4.

[YAO13] Z.-Q. YAO; X.-G. LIANG; L. LUO, & J. HU: A chatter free calibration method for determining cutter runout and cutting force coefficients in ball-end milling. In: *Journal of Materials Processing Technology* 213.9 (Sept. 2013), pp. 1575–1587. ISSN: 0924-0136. DOI: 10.1016/j.jmatprotec.2013.03.023.

[YAO18a] Q. YAO; M. LUO; D. ZHANG, & B. WU: Identification of cutting force coefficients in machining process considering cutter vibration. In: *Mechanical Systems and Signal Processing* 103 (Mar. 2018), pp. 39–59. ISSN: 0888-3270. DOI: 10.1016/j.ymssp.2017.09.038.

[YAO18b] Q. YAO; B. WU; M. LUO, & D. ZHANG: On-line cutting force coefficients identification for bull-end milling process with vibration. In: *Measurement* 125 (Sept. 2018), pp. 243–253. ISSN: 0263-2241. DOI: 10.1016/j.measurement.2018.04.084.

[ZHAN16] D. ZHANG; R. MO; Z. CHANG; H. SUN, & C. LI: A study of computing accuracy of calibrating cutting force coefficients and run-out parameters in flat-end milling. In: *The International Journal of Advanced Manufacturing Technology* 84.1 (Apr. 2016), pp. 621–630. ISSN: 0268-3768, 1433-3015. DOI: 10.1007/s00170-015-7707-9.

[ZHAN17a] X. Zhang; W. Zhang; J. Zhang; B. Pang, & W. Zhao: General Modeling and Calibration Method for Cutting Force Prediction With Flat-End Cutter. In: *Journal of Manufacturing Science and Engineering* 140.2 (Dec. 18, 2017), pp. 021007-1–021007-8. ISSN: 1087-1357. DOI: 10.1115/1.4038371.

[ZHAN17b] Z. Zhang; H. Li; G. Meng; S. Ren, & J. Zhou: A new procedure for the prediction of the cutting forces in peripheral milling. In: *The International Journal of Advanced Manufacturing Technology* 89.5 (Mar. 2017), pp. 1709–1715. ISSN: 0268-3768, 1433-3015. DOI: 10.1007/s00170-016-9186-z.

[ZHAN18a] X. Zhang; J. Zhang; W. Zhang; J. Li, & W. Zhao: A non-contact calibration method for cutter runout with spindle speed dependent effect and analysis of its influence on milling process. In: *Precision Engineering* 51 (Jan. 2018), pp. 280–290. ISSN: 0141-6359. DOI: 10.1016/j.precisioneng.2017.08.020.

[ZHAN18b] Z. Zhang; M. Luo; D. Zhang, & B. Wu: A force-measuring-based approach for feed rate optimization considering the stochasticity of machining allowance. In: *The International Journal of Advanced Manufacturing Technology* 97.5 (July 2018), pp. 2545–2556. ISSN: 0268-3768, 1433-3015. DOI: 10.1007/s00170-018-2127-2.

[ZHU99] R. Zhu; S. G. Kapoor; R. E. DeVor, & S. M. Athavale: Mechanistic Force Models for Chip Control Tools. In: *Journal of Manufacturing Science and Engineering* 121.3 (1999), pp. 408–416. ISSN: 1087-1357. DOI: 10.1115/1.2832696.

[ZUPE06a] U. Zuperl; E. Kiker, & K. Jezernik: Adaptive Force Control in High-Speed Machining by Using a System of Neural Networks. In: *Industrial Electronics, 2006 IEEE International Symposium on* 1 (July 9, 2006), pp. 148–153. DOI: 10.1109/ISIE.2006.295583.

[ZUPE06b] U. Zuperl; F. Cus, & E. Kiker: Intelligent Adaptive Cutting Force Control in End-Milling. In: *TEHNIČKI VJESNIK* 13 (2006), pp. 15–22. ISSN: 1330-3651. URL: https://hrcak.srce.hr/8842.

[ZUPE12a] U. Zuperl; F. Cus, & M. Reibenschuh: Modeling and adaptive force control of milling by using artificial techniques. In:

Journal of Intelligent Manufacturing 23.5 (Oct. 2012), pp. 1805–1815. ISSN: 0956-5515, 1572-8145. DOI: 10.1007/s10845-010-0487-z.

[ZUPE12b] U. ZUPERL & F. CUS: System for off-line feedrate optimization and neural force control in end milling. In: *International Journal of Adaptive Control and Signal Processing* 26.2 (Feb. 2012), pp. 105–123. ISSN: 0890-6327. DOI: 10.1002/acs.1277.

In der Reihe **Ergebnisse aus der Produktionstechnik** zuletzt erschienen:

Greipel, Jonathan:
Grouping of Inspection Characteristics to Reduce Costs of Adaptive Inspection Planning in Variant Production
ISBN: 978-3-98555-157-6
Erscheinungsdatum: 03.05.2023

Troß, Nico:
Mehrskalenmodellierung der thermomechanischen Werkzeugbelastung beim Trockenwälzfräsen
ISBN: 978-3-98555-158-3
Erscheinungsdatum: 19.04.2023

Nicksch, Christoph:
Konfiguration verteilter Large-Scale Metrology Systeme in rekonfigurierbaren Montageumgebungen
ISBN: 978-3-98555-152-1
Erscheinungsdatum: 20.03.2023

Ahmad, Mubarik:
Anregungsgerechte Auslegung und Tolerierung schnelllaufender Getriebe unter Berücksichtigung kurz- und langwelliger Verzahnungsabweichungen
ISBN: 978-3-98555-130-9
Erscheinungsdatum: 06.03.2023

Priwisch, Martin:
Kontaktfreie Bahnlaufregelung zur Herstellung funktionaler Mehrlagensysteme in einem kontinuierlichen Laminierprozess
ISBN: 978-3-98555-139-2
Erscheinungsdatum: 27.02.2023

Welsing, Martin:
Bewertung der ökologischen Nachhaltigkeit von Produktionsnetzwerkkonfigurationen
ISBN: 978-3-98555-155-2
Erscheinungsdatum: 19.04.2023

Beckers, Alexander:
Auslegung wirtschaftlich-technologisch optimierter Fertigungsprozessfolgen mittels prozessübergreifender Modellierung
ISBN: 978-3-98555-154-5
Erscheinungsdatum: 18.04.2023

Gipperich, Marius:
Untersuchung eines Zweistrahlprozesses als Ansatz zur Prozessstabilitätserhöhung des drahtbasierten Laserauftragschweißens
ISBN: 978-3-98555-151-4
Erscheinungsdatum: 15.03.2023

Motschke, Tobias:
Analyse und Modellierung der Dämpfungseigenschaften von Spindellagern
ISBN: 978-3-98555-148-4
Erscheinungsdatum: 06.03.2023

Frye, Maik:
Recommending Data Preprocessing Pipelines for Machine Learning Applications in Production
ISBN: 978-3-98555-147-7
Erscheinungsdatum: 15.02.2023

Edition Wissenschaft
Apprimus

Cremer, Sven:

Controlling der Migration in Produktionsnetzwerken

ISBN: 978-3-98555-145-3
Erscheinungsdatum: 09.02.2023

Ohlert, Marius:

Contact between Abrasive Media and Workpiece in Robot-guided Centrifugal Finishing

ISBN: 978-3-98555- 144-6
Erscheinungsdatum: 07.02.2023

Graberg, Tim:

Operationalisierung von digitalen Schatten im Werkzeugbau

ISBN: 978-3-98555-141-5
Erscheinungsdatum: 06.02.2023

Müller, Tobias:

Modellbildung mittels Symbolischer Regression zur Messunsicherheitsbestimmung komplexer Messprozesse

ISBN: 978-3-98555-146-0
Erscheinungsdatum: 30.01.2023

Kreß, Julian Maximilian:

Erbringung hybrider Leistungsangebote in Wertschöpfungsnetzwerken

ISBN: 978-3-98555-128-6
Erscheinungsdatum: 24.01.2023

Vierschilling, Sebastian Patrick:

Automatisierung der Standortauswahl in der Fabrikplanung

ISBN: 978-3-98555-134-7
Erscheinungsdatum: 09.01.2023

Schmidhuber, Matthias:

Sozio-technische Problemlösungsfähigkeit von PPS-Systemen

ISBN: 978-3-98555- 143-9
Erscheinungsdatum: 07.02.2023

Schwarberg, Thomas:

Bestimmung der Logik von Innovationsideen für technische Systeme

ISBN: 978-3-98555- 142-2
Erscheinungsdatum: 07.02.2023

Nienheysen, Philipp:

Werkstattgerechte Schadensanalyse in CFK-Strukturbauteilen mit optisch georteter Ultraschallmesstechnik

ISBN: 978-3-98555-136-1
Erscheinungsdatum: 30.01.2023

Behrens, Jan:

Selbstkalibrierung von 5-Achs-Bearbeitungszentren mit Dreh-/Schwenktisch

ISBN: 978-3-98555-124-8
Erscheinungsdatum: 24.01.2023

Bickendorf, Philipp:

Tribologische und geometrische Auslegung von additiv hergestellten Werkzeug-einsätzen zur Feinblechumformung

ISBN: 978-3-98555-133-0
Erscheinungsdatum: 24.01.2023

Helbig, Jens:

Typisierung und Auswahl von Referenzstrategien für den Werkzeugeinkauf

ISBN: 978-3-98555-122-4
Erscheinungsdatum: 19.12.2022

Edition Wissenschaft
Apprimus

Zeitfracht Medien GmbH
Ferdinand-Jühlke-Straße 7
99095 Erfurt, Deutschland
produktsicherheit@kolibri360.de